American Book Company
The Standards Experts

MASTERING THE GRADE 5 COMMON CORE IN MATHEMATICS

Erica Day

Colleen Pintozzi

Mary Reagan

Reviewed By:

Lori D. Unruh

AMERICAN BOOK COMPANY

P. O. BOX 2638

WOODSTOCK, GEORGIA 30188-1383

TOLL FREE 1 (888) 264-5877   PHONE (770) 928-2834

TOLL FREE FAX 1 (866) 827-3240

WEB SITE: www.americanbookcompany.com

Acknowledgements

In preparing this book, we would like to acknowledge Eric Field and Mary Stoddard for their contributions in developing graphics, and Alexandra Tatum for her contributions in editing. We would also like to give a special thanks to Jessica DeBord for creating the teaching activities located in the answer key and our many students whose needs and questions inspired us to write this text.

Copyright © 2012
by American Book Company
P.O. Box 2638
Woodstock, GA 30188-1383

ALL RIGHTS RESERVED

The text of this publication, or any part thereof, may not be reproduced or transmitted in any form or by any means, electronic or mechanical, including photocopying, recording, storage in an information retrieval system, or otherwise, without the prior permission of the publisher.

Printed in the United States of America
07/12   01/12

Acknowledgements

Chart of Standards

Standard	Chapter Number
5.OA.1	12
5.OA.2	12
5.OA.3	12
5.NBT.1	1
5.NBT.2	1, 4
5.NBT.3	3
5.NBT.4	3
5.NBT.5	1
5.NBT.6	2
5.NBT.7	4
5.NF.1	5
5.NF.2	5
5.NF.3	7
5.NF.4	6, 9
5.NF.5	6
5.NF.6	6
5.NF.7	7
5.MD.1	8
5.MD.2	8
5.MD.3	10
5.MD.4	10
5.MD.5	10
5.G.1	11
5.G.2	11
5.G.3	9
5.G.4	9

Contents

Acknowledgements		ii
Preface		viii
1	**Whole Numbers**	**1**
1.1	Place Value: Greater Than One (DOK 1)	1
1.2	Digits (DOK 2)	3
1.3	Comparing Whole Numbers (DOK 2)	4
1.4	Adding Whole Numbers (DOK 1)	5
1.5	Subtracting Whole Numbers (DOK 1)	6
1.6	Multiplying Whole Numbers (DOK 2)	7
1.7	Multiplying by Multiples of Ten (DOK 2)	9
1.8	Going Deeper into Whole Numbers (DOK 3)	11
	Chapter 1 Review	12
	Chapter 1 Test	13
2	**Whole Number Division**	**15**
2.1	Divisibility Rules (DOK 1)	15
2.2	Dividing Whole Numbers (DOK 2)	17
2.3	Remainders (DOK 2)	19
2.4	Relationship Between Multiplication and Division (DOK 2)	21
2.5	Modeling Division (DOK 2)	22
2.6	Going Deeper into Whole Number Division (DOK 3)	23
	Chapter 2 Review	24
	Chapter 2 Test	25
3	**Decimals**	**27**
3.1	Place Value: Less Than One (DOK 1, 2)	27
3.2	Reading and Writing Decimal Numbers (DOK 1)	29
3.3	Expanded Form of Decimals (DOK 2)	30
3.4	Estimating Decimals Using Rounding (DOK 2)	31
3.5	Ordering Decimals (DOK 2)	32
3.6	Comparing Decimals (DOK 2)	33
3.7	Going Deeper into Decimals (DOK 3)	34
	Chapter 3 Review	35
	Chapter 3 Test	36

Contents

4 Performing Operations on Decimals — 38
- 4.1 Adding Decimals (DOK 1) — 38
- 4.2 Subtracting Decimals (DOK 1) — 39
- 4.3 Relationship Between Addition and Subtraction of Decimals (DOK 2) — 40
- 4.4 Multiplying Decimals by Multiples of Ten (DOK 2) — 41
- 4.5 Multiplying Decimals by Whole Numbers (DOK 2) — 42
- 4.6 Multiplying Decimals by Decimals (DOK 2) — 43
- 4.7 More Multiplying Decimals by Decimals (DOK 2) — 44
- 4.8 Division of Decimals by Whole Numbers (DOK 2) — 45
- 4.9 Division of Decimals by Decimals (DOK 2) — 46
- 4.10 Modeling Operations on Decimals (DOK 2) — 47
- 4.11 Going Deeper into Performing Operations on Decimals (DOK 3) — 50
- Chapter 4 Review — 51
- Chapter 4 Test — 52

5 Adding and Subtracting Fractions — 54
- 5.1 Fractions (DOK 1) — 54
- 5.2 Simplifying Fractions (DOK 1) — 55
- 5.3 Simplifying Improper Fractions (DOK 1, 2) — 56
- 5.4 Equivalent Fractions (DOK 1, 2) — 58
- 5.5 Adding Fractions (DOK 2) — 60
- 5.6 Subtracting Fractions (DOK 2) — 61
- 5.7 Least Common Multiple (DOK 2) — 62
- 5.8 Adding Mixed Numbers (DOK 2) — 63
- 5.9 Subtracting Mixed Numbers (DOK 2) — 64
- 5.10 Changing Mixed Numbers to Improper Fractions (DOK 1, 2) — 65
- 5.11 Estimating Fractions (DOK 2) — 66
- 5.12 Comparing Fractions (DOK 2) — 68
- 5.13 Fraction Word Problems (DOK 2) — 69
- 5.14 Modeling Fractions (DOK 2) — 70
- 5.15 Going Deeper into Adding and Subtracting Fractions (DOK 3) — 71
- Chapter 5 Review — 72
- Chapter 5 Test — 74

6 Multiplying Fractions — 77
- 6.1 Multiplying Whole Numbers by Fractions (DOK 2) — 77
- 6.2 More Multiplying Whole Numbers by Fractions (DOK 2) — 80

	6.3	Multiplying Fractions with Canceling (DOK 2)	81
	6.4	Multiplying Mixed Numbers (DOK 2)	82
	6.5	Interpreting Multiplication (DOK 2)	83
	6.6	Multiplication Word Problems (DOK 2, 3)	84
	6.7	Modeling Multiplication	85
		Chapter 6 Review	87
		Chapter 6 Test	89
7	**Dividing Fractions**	**91**
	7.1	Interpreting Fractions as Division Problems (DOK 2)	91
	7.2	Reciprocals (DOK 1)	93
	7.3	Dividing Fractions by Whole Numbers (DOK 2)	94
	7.4	Dividing Whole Numbers by Fractions (DOK 2)	95
	7.5	Division Word Problems (DOK 2)	96
	7.6	Modeling Division (DOK 2, 3)	97
		Chapter 7 Review	99
		Chapter 7 Test	100
8	**Measurements and Line Plots**	**102**
	8.1	Customary Measurement (DOK 1)	102
	8.2	Converting Units of Customary Measure (DOK 2)	103
	8.3	Real-World Customary Measurement (DOK 2)	104
	8.4	Metric Measurement (DOK 1)	105
	8.5	Understanding Meters (DOK 1)	105
	8.6	Understanding Liters (DOK 1)	105
	8.7	Understanding Grams (DOK 1)	105
	8.8	Converting Units Within the Metric System (DOK 2)	106
	8.9	Real-World Metric Measurement (DOK 2)	107
	8.10	Line Plots (DOK 2)	108
	8.11	Going Deeper into Measurements and Line Plots (DOK 3)	113
		Chapter 8 Review	114
		Chapter 8 Test	116
9	**Plane Geometry**	**118**
	9.1	Polygons (DOK 1)	118
	9.2	Quadrilaterals and Their Properties (DOK 1)	119
	9.3	Identifying Figures (DOK 1, 2)	120
	9.4	Properties of Shapes (DOK 2)	122

Contents

9.5	Putting Shapes in Categories (DOK 2)	125
9.6	Area of Rectangles (DOK 2)	126
9.7	Going Deeper into Plane Geometry (DOK 3)	128
	Chapter 9 Review	129
	Chapter 9 Test	130

10 Solid Geometry — **132**

10.1	Understanding Volume (DOK 2)	132
10.2	Volume of Rectangular Prisms (DOK 2)	133
10.3	Volume of Cubes (DOK 2)	134
10.4	Volume of Compound Figures (DOK 3)	135
10.5	Real–World Volume Problems (DOK 2, 3)	138
	Chapter 10 Review	139
	Chapter 10 Test	141

11 Introduction to Graphing — **143**

11.1	The Coordinate Grid and Ordered Pairs (DOK 1)	143
11.2	Plotting Points on a Coordinate Grid (DOK 1)	145
11.3	Finding Points From Diagrams (DOK 2, 3)	147
	Chapter 11 Review	149
	Chapter 11 Test	151

12 Algebra and Patterns — **154**

12.1	Algebra Vocabulary (DOK 1)	154
12.2	Understanding Algebra Word Problems (DOK 2)	155
12.3	Equivalent Expressions (DOK 2)	157
12.4	Evaluating Expressions (DOK 2)	159
12.5	Number Patterns (DOK 3)	160
12.6	Pattern Rules (DOK 2, 3)	162
12.7	Graphing Patterns (DOK 2)	164
12.8	Going Deeper into Algebra and Patterns (DOK 3)	166
	Chapter 12 Review	167
	Chapter 12 Test	169

13 How to Write Your Answers — **171**

13.1	Writing Short Answers	171
13.2	Writing Open-Ended Answers	171
	Chapter 13 Review	173

Index — **175**

Preface

Mastering the Grade 5 Common Core in Mathematics will help you review and learn important concepts and skills related to 5th grade mathematics. **The materials in this book are based on the common core standards in mathematics coordinated by the National Governors Association Center for best Practices and the Council of Chief State School Officers. The complete list of standards is located at the beginning of the Answer Key. Each chapter is referenced to the standards.**

This book contains chapters that teach the concepts and skills for 5th Grade. Answers to the tests and exercises are in a separate manual.

Mastering the Grade 5 Common Core in Mathematics includes Depth of Knowledge levels for four content areas based on Norman Webb's Model of interpreting and assigning depth of knowledge levels to both objectives within standards and assessment items for alignment analysis.

The four levels of Depth of Knowledge:

Level 1: Recall and Reproduction (DOK 1)

Questions at this level (DOK 1) include the recall of information such as facts, definitions, or simple procedures. As well as, performing a simple algorithm, or carrying out a one-step, well-defined, and straight forward procedure. A few example DOK level 1 questions are listed below.

· State the associative property of multiplication

· Identify the divisor in a problem

· Measure the perimeter of a figure

· Calculate $4.3 + 8.5$

Level 2: Skills and Concepts/Basic Reasoning (DOK 2)

DOK level 2 questions involve some mental processing beyond a habitual response. It requires students to make some decisions as to how to approach the problem. As well as being able to classify, organize, estimate, make observations, collect, display, and compare data. A few example DOK 2 level questions are listed below.

· Interpret the bar graph to answer questions about a given population

· Classify different types of polygons based upon their characteristics

· Compare two sets of data using their measures of central tendencies

· Extend an algebraic pattern

Level 3: Strategic Thinking/Complex Reasoning (DOK 3)

This level (DOK 3) includes problems that require reasoning, planning, using evidence, and higher levels of thinking beyond what was required in DOK levels one and two. This level requires students to explain their thinking, and cognitive demands are more complex and abstract. DOK 3 demands that students use reasoning skills to draw conclusions from observations and make conjectures. Some examples of level 3 DOK questions are listed below.

· Explain how you can determine if two triangles are similar

· Formulate an expression to determine the next few terms in a pattern

· Construct a survey and analyze the results to determine the most popular movie genre

Level 4: Extended Thinking/Reasoning (DOK 4)

DOK level 4 questions include things such as: complex reasoning, planning, and developing. Student thinking will most likely take place over an extended period of time, and that will include taking into consideration a number of variables. Students should be required to make several connections and relate ideas within the content area or among other content areas. By selecting one approach among many alternatives on how a situation should be solved. At this level students will be expected to design and conduct their own experiments, make connections between findings and relate them to concepts and phenomena together. A few example problems are presented below:

· Explore real world phenomena of Cartesian plans and create a report to present your findings

· Connect your knowledge of integers to the plate tectonics of Earth

· Analyze common game pieces (i.e. dice, spinners, etc.) to determine their fairness based upon what you know about probability of events by designing and carrying out your own experiment

ABOUT THE AUTHORS

Erica Day has a Bachelor of Science degree in Mathematics and is working on a Master of Science Degree in Mathematics. She graduated with high honors from Kennesaw State University in Kennesaw, Georgia. She has also tutored all levels of mathematics, ranging from high school algebra and geometry to university-level statistics, calculus, and linear algebra. She is currently writing and editing mathematics books for American Book Company, where she has coauthored numerous books, such as ***Passing the Georgia Algebra I End of Course, Passing the Georgia High School Graduation Test in Mathematics***, and ***Passing the New Jersey HSPA in Mathematics***, to help students pass graduation and end of course/grade exams.

Colleen Pintozzi has taught mathematics at the elementary school, middle school, junior high, senior high, and adult level for 22 years. She holds a B.S. degree from Wright State University in Dayton, Ohio and has done graduate work at Wright State University, Duke University, and the University of North Carolina at Chapel Hill. She is the author of many mathematics books including such best-sellers as ***Basics Made Easy: Mathematics Review, Passing the New Alabama Graduation Exam in Mathematics, Passing the Louisiana LEAP 21 GEE, Passing the Indiana ISTEP + GQE in Mathematics, Passing the Minnesota Basic Standards in Mathematics***, and ***Passing the Nevada High School Proficiency Exam in Mathematics***.

ABOUT THE REVIEWER

Lori D Unruh has a Bachelor of Science Degree in Elementary Education with an emphasis in elementary mathematics from Northwestern Oklahoma State University. Ms. Unruh has also completed her Master of Education Degree in the area of elementary and secondary guidance counseling from Southwestern Oklahoma State University. She has been an educator for 25 years in the states of Oklahoma, Texas, and Kansas. Ms. Unruh started the Grant County Alternative Education Academy and was the director and counselor for the academy for two years. She is currently a K-12 grade guidance counselor, art instructor, and district testing coordinator for the Pond Creek-Hunter School District.

Preface

This book is interactive!

Augmented Reality is an exciting technology that allows you to interact with printed material from Android and iOS smartphones and tablets. We have implemented this technology in this book to enhance your students' learning experience in a more visual way!

Use your smartphone or tablet to scan the QR Code to the left and download the AR app for this book.

When you see the icon on the right throughout the book, refer back to this page to experience your AR content.

Refer to the "About AR" on page xi!

American Book Company

The Standards Experts

AUGMENTED REALITY

xi

Preface

To learn how to use ABC's Augmented Reality, scan the QR Code to the left with your smartphone or tablet to watch a tutorial video.

You can also visit our web site at **americanbookcompany.com/AugmentedReality** to view the tutorial video.

Below is a list of the locations of the AR content exercises in this book.

Examples

Chapter 1
Page 9 – Example 6

Chapter 2
Page 19 – Example 4

Chapter 3
Page 29 – Example 3

Chapter 4
Page 41 - Example 5

Chapter 5
Page 64 – Example 12

Chapter 6
Page 82 – Example 6

Chapter 7
Page 95 – Example 6

Chapter 8
Page 106 – Example 1

Chapter 9
Page 126 – Example 2

Chapter 10
Page 133 – Example 2

Chapter 11
Page 145 – Example 3

Chapter 12
Page 159 – Example 2

Activities

1) River/Rabbit Game (Rounding, Place Value)
2) Car Game (Multiplication, Addition, Place Value, Powers of Ten)

Chapter 1
Whole Numbers

This chapter covers the following CC 5 standards:

Number and Operations in Base 10	5.NBT.1, 5.NBT.2, 5.NBT.5

1.1 Place Value: Greater Than One (DOK 1)

Place Value: The value of a digit based upon its place, within the number. For example, in the chart below, the number 7 has two very different values in the number 7,867,413:

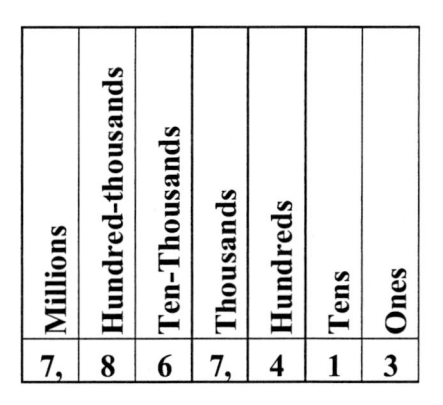

Millions	Hundred-thousands	Ten-Thousands	Thousands	Hundreds	Tens	Ones
7,	8	6	7,	4	1	3

As we progress from right to left, we see that each column represents 10 times more than the column before it:

As we progress from left to right, we see that each column represents one tenth of the value of the column before it:

Ones: 1
Tens: $10 \times 1 = 10$
Hundreds: $10 \times 10 = 100$
Thousands: $10 \times 100 = 1000$
Ten-thousands: $10 \times 1000 = 10,000$
Hundred-thousands: $10 \times 10,000 = 100,000$
Millions: $10 \times 100,000 = 1,000,000$

Millions: $1,000,000$
Hundred-thousands: $1,000,000 \div 10 = 100,000$
Ten-thousands: $100,000 \div 10 = 10,000$
Thousands: $10,000 \div 10 = 1,000$
Hundreds: $1,000 \div 10 = 100$
Tens: $100 \div 10 = 10$
Ones: $10 \div 10 = 1$

Copyright © American Book Company

Chapter 1   Whole Numbers

Using the same number: **7,867,413**
As we progress from right to left:

Ones:	$3 \times 1 = 3$
Tens:	$1 \times 10 = 10$
Hundreds:	$4 \times 100 = 400$
Thousands:	$7 \times 1,000 = 7,000$
Ten-thousands:	$6 \times 10,000 = 60,000$
Hundred-thousands:	$8 \times 100,000 = 800,000$
Millions:	$7 \times 1,000,000 = 7,000,000$

Let's write this number in expanded form to clearly identify the value of each number.

The number written in **standard** form: $7,867,413$

The number written in **expanded** form: $7,000,000 + 800,000 + 60,000 + 7,000 + 400 + 10 + 3 = 7,867,413$

We read it as follows: seven million, eight hundred sixty-seven thousand, four hundred thirteen.

Write the value of the number 4 in word form. The first one has been done for you. (DOK 1)

1. 14,167   Four thousand
2. 340,625 _____
3. 64 _____
4. 4,678,123 _____
5. 46 _____
6. 74 _____
7. 413 _____
8. 14 _____
9. 874,115 _____

Use this number to answer the following questions: 3,746,028 (DOK 1)
Which number is in the

10. Tens place? _____
11. Hundred-thousands place? _____
12. Millions place? _____
13. Ones place? _____
14. Thousands place? _____
15. Hundreds place? _____
16. Ten-thousands place? _____

Write the value of the underlined digit. (DOK 1)

17. 6<u>6</u>0 _____
18. <u>2</u>0,547 _____
19. 8,<u>7</u>98,620 _____
20. 114,0<u>6</u>1 _____
21. <u>8</u>90 _____
22. <u>3</u>,132,894 _____

1.2 Digits (DOK 2)

A **digit** is a number in a given place value. For instance, for the number 104, there are three digits. The 1 is the hundreds place digit, the 0 is the tens place digit, and the 4 is the ones place digit.

Example 1: Find the values of digits in a number.

Going from right to left: Each digit is worth ten times more than the digit to its right.

> The **tens** place is worth 10 times more than the ones place.
> The **hundreds** place is worth 10 times more than the tens place.
> The **thousands** place is worth 10 times more than the hundreds place.
> The **ten-thousands** place is worth 10 times more than the thousands place.
> The **hundred-thousands** place is worth 10 times more than the ten-thousands place.
> The **millions** place is worth 10 times more than the hundred-thousands place.

Going from left to right: Each digit is worth $\frac{1}{10}$ of the digit to its left.

> The **hundred-thousands** place is worth $\frac{1}{10}$ times the millions place.
> The **ten-thousands** place is worth $\frac{1}{10}$ times the hundred-thousands place.
> The **thousands** place is worth $\frac{1}{10}$ times the ten-thousands place.
> The **hundreds** place is worth $\frac{1}{10}$ times the thousands place.
> The **tens** place is worth $\frac{1}{10}$ times the hundreds place.
> The **ones** place is worth $\frac{1}{10}$ times the tens place.

Compare the digits by finding if one number should be multiplied by 10 or $\frac{1}{10}$ of the other. Questions 3 and 4 are done for you. (DOK 2)

1. How many digits does 554,389 have?

2. How many digits does 7,000,001 have?

3. Compare the hundreds place to the thousands place.
   Answer: The hundreds place is $\frac{1}{10}$ of the thousands place.

4. Compare the tens place to the ones place.
   Answer: The tens place is 10 times more than the ones place.

5. Compare the millions place to the hundred-thousands place.

6. Compare the ten-thousands place to the thousands place.

7. Compare the hundreds place to the tens place.

8. Compare 100,000 to 1,000,000.

9. Compare 100,000 to 10,000.

10. Compare 7,000 to 70,000.

11. Compare the ten-thousands place to the hundred-thousands place.

Chapter 1   Whole Numbers

1.3   Comparing Whole Numbers (DOK 2)

To compare whole numbers is to decide which is smaller and which is larger. The symbol < means less than. The symbol > means greater than. Either symbol will always point to the smaller number. For instance, you know that three is less than five. $3 < 5$ or $5 > 3$. Notice how the symbols point to the smaller number.

In word form: $3 < 5$ is "three is less than five" $5 > 3$ is "five is greater than three"

Example 2:   Compare the digits of the numbers 1,329 and 1,340. Which number is bigger?

Step 1:   Line up the numbers (if the numbers have the same number of digits).   1,329
1,340

Step 2:   Compare the first digit (on the left) in each number: **1**,329
1,340

They are the same. Go to the next digit in each number.

Step 3:   Look at the second digit in each number: 1,**3**29
1,**3**40

Again, the digits are the same. Go to the next digit in each number.

Step 4:   Look at the third digit in each number: 1,3**2**9
1,3**4**0

These digits are not the same. Which is bigger? The 4 is bigger than the 2, so 1,340 is bigger than 1,329.

Answer: $1,340 > 1,329$.

Compare the pairs of numbers below and chose the correct symbol, < or >. (DOK 2)

1. 232 _____ 293

2. 11,001 _____ 11,010

3. 49 _____ 48

4. 503,002 _____ 203,005

5. 1,117 _____ 1,890

6. 40,200 _____ 42,000

7. 6,000,000 _____ 600,000

8. 888,088 _____ 888,880

9. 33,013 _____ 33,301

10. 442,255 _____ 422,244

11. 67,677 _____ 76,677

12. 911,119 _____ 199,991

13. 327,187 _____ 372,178

14. 55,015 _____ 550,015

1.4 Adding Whole Numbers (DOK 1)

Example 3: Find $302 + 54 + 712 + 9$.

Step 1: Remember when you add to arrange the numbers in columns with the ones digits at the right.

$$\begin{array}{r} 302 \\ 54 \\ 712 \\ +9 \\ \hline \end{array}$$

Step 2: Start at the right and add each column. Remember to carry when necessary.

$$\begin{array}{r} 1 \\ 302 \\ 54 \\ 712 \\ +9 \\ \hline 1,077 \end{array}$$

Find the sum. (DOK 1)

1. $18 + 24 + 157$

2. $2,458 + 5,011$

3. $4,005 + 1,342$

4. $386 + 54 + 3$

5. $4,057 + 21 + 219$

6. $2,465 + 486$

7. The total of 9 and 104

8. 94 more than 541

9. 784 increased by 51

10. 18 more than 149

11. 5 more than 557

12. 102 added to 73

13. 298 increased by 25

14. 541 plus 402

15. $12 + 454 + 3 + 97$

16. The sum of 308 and 52

17. $5,442,001 + 1,339,100$

18. $3,780,395 + 4,118,234$

Chapter 1   Whole Numbers

1.5   Subtracting Whole Numbers (DOK 1)

Example 4:   Find $1,006 - 568$.

Step 1:   Remember when you subtract to arrange the numbers in columns with the ones digits at the right.

$$\begin{array}{r} 1006 \\ -568 \\ \hline \end{array}$$

Step 2:   Start at the right, and subtract each column. Remember to borrow when necessary.

$$\begin{array}{r} \overset{9\;9}{\cancel{1\,0\,0}}{}^{1}6 \\ -\;\;5\,6\,8 \\ \hline 4\,3\,8 \end{array}$$ ← Borrow 1 from the 100, making it 99.

Note: When you see "less than" in a problem, the second number becomes the top number when you set up the problem.

Find the difference. (DOK 1)

1. $541 - 35$ _____

2. $6,007 - 279$ _____

3. $694 - 287$ _____

4. $902 - 471$ _____

5. $500 - 376$ _____

6. $1,047 - 483$ _____

7. 14 less than 607 _____

8. 881 decreased by 354 _____

9. The difference between 384 and 29 _____

10. 560 decreased by 125 _____

11. 43 less than 752 _____

12. 74 less than $1,093$ _____

13. 96 less than 704 _____

14. 327 less than $1,002$ _____

15. The difference between 273 and 55 _____

16. The difference between $2,849$ and 756 ___

17. $8,772,455 - 6,228,417$ _____

18. $3,228,536 - 2,762,844$ _____

1.6 Multiplying Whole Numbers (DOK 2)

Example 5: Multiply: 256×73.

Step 1: Line up the ones digits. Multiply 256×3.

$$\begin{array}{r} \overset{1\ 1}{2\ 5\ 6} \\ \times\ 7\boxed{3} \\ \hline 7\ 6\ 8 \end{array}$$

$\begin{cases} 3 \times 6 = 18, \text{ write 8 and carry the one} \\ 3 \times 5 = 15, \text{ add the 1 that was carried to get 16,} \\ \qquad\text{write 6 and carry the one} \\ 3 \times 2 = 6, \text{ add the 1 that was carried to get 7,} \\ \qquad\text{write 7} \end{cases}$

Step 2: Multiply 256×7. Remember to put the partial product one place to the left under the tens column. Then add.

$$\begin{array}{r} \overset{3\ 4}{2\ 5\ 6} \\ \times\ \boxed{7}3 \\ \hline 7\ 6\ 8 \\ 1\ 7\ 9\ 2 \\ \hline 1\ 8{,}6\ 8\ 8 \end{array}$$

$\begin{cases} 7 \times 6 = 42, \text{ write 2 and carry the 4} \\ 7 \times 5 = 35, \text{ add the 4 that was carried to get 39,} \\ \qquad\text{write 9 and carry the 3} \\ 7 \times 2 = 14, \text{ add the 3 that was carried to get 17,} \\ \qquad\text{write 17} \end{cases}$

Add

The answer is 18,688.

Multiply. Show your work. (DOK 2)

1.	258 × 72	6.	324 × 19	11.	581 × 25	16.	456 × 47
2.	742 × 44	7.	921 × 23	12.	827 × 56	17.	743 × 65
3.	785 × 32	8.	454 × 56	13.	942 × 24	18.	527 × 38
4.	679 × 36	9.	156 × 95	14.	247 × 84	19.	524 × 39
5.	841 × 27	10.	765 × 94	15.	468 × 43	20.	682 × 64

Chapter 1   Whole Numbers

21. 2,162,334
   ×   4

22. 544,455
   ×   5

23. 3,178,421
   ×   3

24. 16,599
   ×   24

25. 54,877
   ×   13

26. 4,289,117
   ×   2

27. 373,566
   ×   7

28. 92,102
   ×   37

29. 1,922,210
   ×   5

30. 462,152
   ×   11

31. 32,658
   ×   74

32. 80,443
   ×   52

33. 2,642,333
   ×   2

34. 890,247
   ×   9

35. 1,533
   ×   482

36. 7,230
   × 1,327

37. 6,477
   × 1,500

38. 3,221,926
   ×   2

39. 1,990,004
   ×   3

40. 485,115
   ×   17

41. 312,705
   ×   23

1.7 Multiplying by Multiples of Ten (DOK 2)

Multiples of ten are any number that ends in a zero, such as 50, 234, 880, and 760. Multiply each number by a another factor of 10 to find the multiples of a number.

Example 6: Multiply 479×100.

Step 1: Count the number of 0's in the multiple of 10. There are two 0's in 100.

Step 2: Add two zeros to the other factor in the problem and write the answer.
$479 \times 100 = 47900$

Step 3: Put the commas in the appropriate places: $47,900$.

Answer: $47,900$

Refer to the "About AR" on page xi!

Find the product. (DOK 2)

1. 27×100 _____
2. $356 \times 1,000$ _____
3. $471 \times 100,000$ _____
4. $3,714 \times 100$ _____
5. $2,642 \times 1,000$ _____
6. $1,261 \times 100,000$ _____
7. 39×100 _____
8. $42 \times 1,000$ _____
9. $33 \times 100,000$ _____

Multiplying numbers by a multiple of ten is also called multiplying by a power of ten. The powers of ten are abbreviated by use of an exponent. In the number 10^3, the 10 is the base number and the 3 is the exponent. The exponent tells you how many times to multiply the base number by itself. $10^3 = 10 \times 10 \times 10 = 1,000$. Note that $1,000$ has three zeros in it, the same as the number in the exponent.

$10^2 = 10 \times 10 = 100$	100 has 2 zeros
$10^3 = 10 \times 10 \times 10 = 1,000$	$1,000$ has 3 zeros
$10^4 = 10 \times 10 \times 10 \times 10 = 10,000$	$10,000$ has 4 zeros
$10^5 = 10 \times 10 \times 10 \times 10 \times 10 = 100,000$	$100,000$ has 5 zeros
$10^6 = 10 \times 10 \times 10 \times 10 \times 10 \times 10 = 1,000,000$	$1,000,000$ has 6 zeros

When multiplying numbers by a power of ten, take the first factor and add to it zeros equalling the same number as the exponent.

Example 7: Multiply 84×10^5.

Step 1: Start the product with the factor without the exponent: 84.

Step 2: Add to 84 the same number of zeros as the number in the exponent. In this case, the exponent is 5, so add 5 zeros.
8400000

Step 3: Put the commas in the appropriate places: $8,400,000$.

Answer: $8,400,000$

Chapter 1   Whole Numbers

Multiply. (DOK 2)

1. 17×10^3 _____
2. 324×10^4 _____
3. 8×10^2 _____
4. $1,923 \times 10^3$ _____
5. 624×10^2 _____

6. 55×10^5 _____
7. 3×10^6 _____
8. 48×10^3 _____
9. 95×10^3 _____
10. 112×10^4 _____

11. 7×10^5 _____
12. $7,651 \times 10^3$ _____
13. 286×10^2 _____
14. 64×10^5 _____
15. 9×10^6 _____

Numbers may be broken down into a factor times a power of ten.

Example 8: Break down $3,051,000$ into a factor times a power of ten.

Step 1: Take the first digits of the number until there are only zeros left.
The number $3,051,000$ starts with the digits 3051, with the commas removed.

Step 2: Multiply this number by the power of ten. The power of ten will equal the number of zeros. In this case there are 3 zeros after the number.
3051×10^3.

Step 3: Put commas back in the first factor where appropriate: $3,051$.

Answer: $3,051,000 = 3,051 \times 10^3$

Break down the numbers below into a factor times a power of ten. (DOK 2)

1. $82,400$ _____
2. $1,623,000$ _____
3. $8,000$ _____
4. $170,000$ _____
5. $5,000,000$ _____

6. $960,000$ _____
7. $2,700$ _____
8. $8,200,000$ _____
9. 300 _____
10. $465,000$ _____

11. $8,000,000$ _____
12. $706,000$ _____
13. $110,000$ _____
14. $5,100,000$ _____
15. $9,000,000$ _____

1.8 Going Deeper into Whole Numbers (DOK 3)

Solve the multi-step word problems below. Show your work for each step. (DOK 3)

1. John was asked to show $9,230,000$ expressed as a power of ten. He wrote $9,230 \times 10^4$. His answer is incorrect. Find the correct answer and explain how you found your answer.

2. There are 24 electronic games in each case of electronic games. Find the number of electronic games in $15,000$ cases and express your answer using a number multiplied by a power of ten.

3. A small city has 7 schools. The number of students attending the schools are shown in the table below.

School	Population
Everest Elementary	1,240
Mountain Elementary	1,410
Valley Elementary	1,330
Meadow Middle School	1,985
Pine Middle School	2,115
Oak View High School	2,095
Maple High School	2,215

Part 1: How many school age children are there in the city?

Part 2: If the population of the elementary, middle, and high school age children represent about one third of the total population of the city, estimate the total population of the city to the nearest hundred. Express your answer using a number multiplied by a power of ten.

Part 3: Comparing the number of elementary school students to the number of high school students, would you conclude that the population of this city is increasing or decreasing?

4. The chart below shows the number of bicycle wheels manufactured in four months at a factory.

Month	Number of Wheels
April	22,500
May	24,000
June	23,474
July	24,150

Part 1: How many wheels were manufactured in the months of April through July?

Part 2: If the four months represent one third of a years production of bicycle wheels, estimate to the nearest thousand, the number of wheels that will be manufactured in one year. Express your answer using a number multiplied by a power of ten.

5. Find the number of minutes in 100 days. Show your work and express your answer using a number multiplied by a power of ten.

Compare the following numbers using $<$, $>$, and $=$. Show your work by writing out the products of each side of the comparison.

6. 46×10^5 _____ 64×10^4

7. 1101×10^3 _____ 1110×10^5

8. 7×10^6 _____ 7×10^4

9. 223×10^2 _____ 322×10^2

10. 100×10^3 _____ 10×10^4

11. $1,000 \times 10^4$ _____ 100×10^6

12. 99×10^6 _____ 99×10^5

Chapter 1   Whole Numbers

Chapter 1 Review

Give the place value of the digit 8 in each number below. (DOK 1)

1. 4,781,003 ____
2. 628 ____
3. 108,056 ____
4. 8,346,213 ____
5. 284 ____
6. 1,823,019 ____
7. 671,358 ____
8. 3,482 ____

Compare the following numbers using <, >, and =. Show your work by writing out the products of each side of the comparison. (DOK 3)

9. 652×10^3 ____ 672×10^2
10. 41×10^5 ____ 401×10^4
11. $1,001 \times 10^2$ ____ 101×10^3
12. $6,812 \times 10^3$ ____ $6,812 \times 10^2$

Add or subtract as indicated. (DOK 1)

13. 48,113 − 24,371
14. 1,228 + 541
15. 971,525 − 32,567
16. 28,671 + 17,113
17. 61,791 − 566
18. 1,417,622 + 238,445
19. 889 − 672
20. 11,118 + 88,881

Multiply. (DOK 2)

21. 2,004,100 × 4
22. 327 × 16
23. 341,337 × 3
24. 645 × 22

Multiply. (DOK 2)

25. 16×10^3 ____
26. 7×10^6 ____
27. 841×10^4 ____
28. 22×10^2 ____

Break down the numbers below into a factor times a power of ten. (DOK 2)

29. 5,600 ____
30. 8,200,000 ____
31. 32,000 ____
32. 140,000 ____

Compare the digits by finding if one number should be multiplied by 10 or $\frac{1}{10}$ of the other. (DOK 2)

33. Compare the millions place to the hundred-thousands place.
34. Compare the ten-thousands place to the thousands place.
35. Compare the hundreds place to the thousands place.
36. Compare the thousands place to the hundreds place.

Chapter 1 Test

Choose the correct answer.

1 Which place value is the 6 in 46,542?

　A tens
　B hundreds
　C thousands
　D ten-thousands

(DOK 1)

2 Which place value is the 3 in 7,354,116?

　A millions
　B hundred-thousands
　C hundred-thousandths
　D ten-thousands

(DOK 1)

3 Which place value is the 1 in 500,751?

　A thousands
　B ten-thousands
　C hundred-thousands
　D ones

(DOK 1)

4 Which number is less than 87,037?

　A 87,370
　B 83,703
　C 87,073
　D 87,730

(DOK 2)

5 Which of these comparisons is true?

　A $87 < 78$
　B $78 > 87$
　C $77 > 88$
　D $88 > 78$

(DOK 2)

6 Which of these comparisons is not true?

　A $33 > 32$
　B $32 < 33$
　C $32 > 33$
　D $32 > 23$

(DOK 2)

7 Add: $42,781 + 6,895$

　A 49,677
　B 48,676
　C 49,676
　D 48,677

(DOK 1)

8 Solve: 5,113,120 plus 4,007,013

　A 9,120,133
　B 9,120,113
　C 9,117,133
　D 9,119,113

(DOK 1)

9 Subtract: $371,016 - 72,008$

　A 298,008
　B 298,018
　C 299,018
　D 299,008

(DOK 1)

10 Solve: 845 less 322

　A 521
　B 523
　C 513
　D 533

(DOK 1)

11 Multiply: 954×11

　A 9,541
　B 10,541
　C 10,494
　D 10,495

(DOK 2)

12 Solve: 1,842,316 times 2

　A 3,674,362
　B 3,684,368
　C 3,684,632
　D 3,674,368

(DOK 2)

Copyright © American Book Company

Chapter 1 Whole Numbers

13 Which number is equal to 37×10^4?

  A 37,000
  B 370,000
  C 370,001
  D 3,700,000

(DOK 2)

14 Which number is equal to 142×10^2?

  A 14,200
  B 1,420
  C 142,000
  D 1,420,000

(DOK 2)

15 Break down 7,640,000 into a factor times a power of ten.

  A $7,640 \times 10^4$
  B 764×10^3
  C $7,640 \times 10^5$
  D 764×10^4

(DOK 2)

16 Breakdown 870,000 into a factor times a power of ten.

  A 870×10^6
  B 87×10^3
  C 87×10^4
  D 87×10^2

(DOK 2)

17 Which number is equal to 444×10^3?

  A 444,000
  B 440,000
  C 444,400
  D 4,440,000

(DOK 2)

18 Breakdown 500 into a factor times a power of ten.

  A 50×10^2
  B 5×10^2
  C 5×10^3
  D 50×10^3

(DOK 2)

19 Which statement is true?

  A The hundreds place is 10 times more than the thousands place.
  B The hundreds place is 10 times more than the ones place.
  C The hundreds place is $\frac{1}{10}$ of the tens place.
  D The hundreds place is $\frac{1}{10}$ of the thousands place.

(DOK 2)

20 Which statement is true?

  A The millions place is 10 times more than the hundred-thousands place.
  B The millions place is $\frac{1}{10}$ of the hundred-thousands place.
  C The hundred-thousands place is 10 times more than the millions place.
  D The hundred-thousands place is $\frac{1}{10}$ of the thousands place.

(DOK 2)

21 There are 52 weeks in one year. How many weeks are there in 1,500 years? Express your answer using a number multiplied by a power of ten.

  A 52×10^3
  B 52×10^2
  C 78×10^3
  D 78×10^2

(DOK 3)

22 Which of the following comparisons are incorrect?

  A $99 \times 10^6 > 99 \times 10^5$
  B $99 \times 10^4 > 99 \times 10^4$
  C $99 \times 10^3 < 99 \times 10^5$
  D $909 \times 10^5 > 909 \times 10^3$

(DOK 3)

Chapter 2
Whole Number Division

This chapter covers the following CC 5 standard:

| Number and Operations in Base 10 | 5.NBT.6 |

2.1   Divisibility Rules (DOK 1)

Divisibility - The ability of a number to be divided evenly, without a remainder.

Example 1:   Is 12 divisible by 6?

Divide 6 into 12 to determine if the quotient has a remainder.

$12 \div 6 = 2 \text{ R}0$

No remainder = Divisibility

This quotient does not have a remainder. Therefore, 12 is divisible by 6.

Example 2:   Is 7 divisible by 3?

Divide 3 into 7 to determine if the quotient has a remainder.

$7 \div 3 = 2 \text{ R}1$

This quotient does have a remainder.

Therefore, 7 is not divisible by 3.

Chapter 2 Whole Number Division

Use the chart below to help you quickly determine if a particular number is divisible by 2, 3, 5, and/or 10.

Divisibility Rules Chart

A number is divisible by ...	if...
2	the last digit (ones place) is zero or an even number.
3	the sum of all the digits is divisible by 3. **Example:** $657 = 6 + 5 + 7 = 18$; $18 \div 3 = 6$. Therefore, 657 is divisible by 3.
5	the last digit (ones place) is a 0 or 5. **Example:** $1,125 \div 5 = 225$
6	the number is divisible by 2 <u>and</u> 3.
10	the last digit is zero. (Example $20 \div 10 = 2$)

Review the numbers below. Write whether they are divisible by 2, 3, 5, 6, or 10. You may have more than one answer. (DOK 1)

1. 27 _____
2. 45 _____
3. 50 _____
4. 62 _____
5. 186 _____
6. 486 _____
7. 1,255 _____
8. 64 _____
9. 820 _____
10. 2,170 _____
11. 4,008 _____
12. 63 _____
13. 8,153,200 _____
14. 66,333 _____
15. 18,425 _____
16. 18 _____
17. 24,000 _____
18. 1,855 _____
19. 12 _____
20. 6,984,236 _____
21. 1,621,750 _____
22. 55 _____
23. 112 _____
24. 5,675 _____

2.2 Dividing Whole Numbers (DOK 2)

Dividend ÷ Divisor = Quotient

Example 3: Divide: $364 \div 7$

Step 1: Rewrite the problem using the symbol $\overline{)}$.

Step 2: Look at the first number large enough to be divided by 7. The number 3 cannot be divided by 7. The number 36 <u>can</u> be divided by 7. Divide 36 by 7. Multiply 7×5, and subtract.

$$\begin{array}{r} 5 \\ 7{\overline{\smash{)}364}} \\ \underline{35} \\ 1 \end{array}$$

Step 3: You will notice you cannot divide 1 by 7. You must bring down the 4. Divide 14 by 7 and subtract.

$$\begin{array}{r} 52 \\ 7{\overline{\smash{)}364}} \\ \underline{35} \\ 14 \\ \underline{14} \\ 0 \end{array}$$

Answer: The answer is 52.

Divide. (DOK 2)

1. $550 \div 5 =$ ____
2. $249 \div 3 =$ ____
3. $416 \div 4 =$ ____
4. $642 \div 2 =$ ____
5. $279 \div 9 =$ ____
6. $200 \div 8 =$ ____
7. $427 \div 7 =$ ____
8. $162 \div 3 =$ ____
9. $685 \div 5 =$ ____
10. $972 \div 9 =$ ____
11. $667 \div 29 =$ ____
12. $714 \div 17 =$ ____
13. $572 \div 22 =$ ____
14. $270 \div 18 =$ ____
15. $910 \div 35 =$ ____
16. $936 \div 52 =$ ____
17. $968 \div 44 =$ ____
18. $294 \div 21 =$ ____
19. $504 \div 12 =$ ____
20. $663 \div 51 =$ ____
21. $476 \div 17 =$ ____

Chapter 2   Whole Number Division

22. $1,320 \div 15 =$ ____

23. $4,248 \div 18 =$ ____

24. $8,376 \div 12 =$ ____

25. $9,462 \div 83 =$ ____

26. $6,875 \div 55 =$ ____

27. $1,809 \div 27 =$ ____

28. $4,704 \div 14 =$ ____

29. $5,010 \div 10 =$ ____

30. $1,218 \div 29 =$ ____

31. $9,386 \div 38 =$ ____

32. $4,873 \div 11 =$ ____

33. $5,586 \div 42 =$ ____

34. $8,534 \div 17 =$ ____

35. $8,316 \div 77 =$ ____

36. $9,191 \div 91 =$ ____

37. $5,365 \div 37 =$ ____

38. $9,256 \div 52 =$ ____

39. $2,875 \div 25 =$ ____

40. $5,520 \div 16 =$ ____

41. $4,536 \div 28 =$ ____

42. $4,212 \div 36 =$ ____

43. $8,466 \div 83 =$ ____

44. $7,533 \div 31 =$ ____

45. $8,333 \div 13 =$ ____

46. $7,429 \div 23 =$ ____

47. $7,954 \div 41 =$ ____

48. $9,557 \div 19 =$ ____

49. $9,425 \div 25 =$ ____

50. $8,118 \div 82 =$ ____

51. $8,366 \div 94 =$ ____

52. $7,360 \div 64 =$ ____

53. $8,640 \div 27 =$ ____

54. $8,930 \div 94 =$ ____

55. $2,541 \div 33 =$ ____

56. $9,800 \div 40 =$ ____

57. $8,712 \div 88 =$ ____

58. $3,654 \div 42 =$ ____

59. $9,810 \div 15 =$ ____

60. $6,058 \div 13 =$ ____

61. $1,541 \div 23 =$ ____

62. $3,392 \div 53 =$ ____

63. $2,262 \div 58 =$ ____

64. $5,530 \div 14 =$ ____

65. $1,988 \div 71 =$ ____

66. $7,808 \div 16 =$ ____

67. $6,615 \div 63 =$ ____

68. $1,372 \div 49 =$ ____

2.3 Remainders (DOK 2)

The **remainder** is the part that is left over after long division.

Example 4: Divide $97 \div 4$

$$\begin{array}{r} 24\text{ R1} \\ 4\overline{)97} \\ -8\ \ (4\times 2)\\ \hline 17 \\ -16\ \ (4\times 4)\\ \hline 1 \end{array}$$

We cannot divide 4 into 1, so it is listed as the remainder. We do this by inserting a capital R and the remaining amount. (1 in this example) R1.

Check:

$$\begin{array}{r} 24 \\ \times\ 4 \\ \hline 96 \end{array} + 1 = 97\ \checkmark\checkmark$$

Refer to the "About AR" on page xi!

Sometimes when your answer contains a remainder, you have to round up or down to solve the problem correctly.

Example 5: Dawn has 22 pictures to put in her album. She can fit 4 on each page.

How many pages will she need to fit all 22 pictures in her album?

$$\begin{array}{r} 5\text{ R2} \\ 4\overline{)22} \\ -20 \\ \hline 2 \end{array}$$

In this example, she will round up to 6 pages. 5 pages will have 4 pictures. 1 page will have the 2 remaining pictures on it.

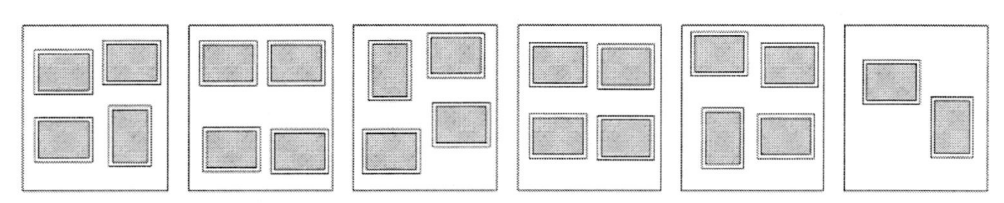

4 pictures + 4 pictures + 4 pictures + 4 pictures + 4 pictures + 2 pictures
= 22 pictures

Answer: 6 pages

Chapter 2   Whole Number Division

Example 6: Todd has $18. He wants to buy as many water bottles as he can. If each bottle costs $4, how many can he buy?

$$\begin{array}{r} 4\text{ R}2 \\ 4\overline{)18} \\ -16 \\ \hline 2 \end{array}$$

Todd can buy 4 bottles. He will have $2 left and that won't be enough to buy another water bottle. The answer is 4.

Solve the problems. (DOK 2)

1. Olivia has 15 flowers. She was asked to place 2 on each table. How many tables can she put 2 flowers on, and how many flowers are leftover? _____

2. Paul has 22 books. He is to put 3 books in each box, how many boxes will he need for <u>all</u> the books? _____

3. Craig's class is going on a field trip. There are 23 students in his class. Each car can carry 4 students. How many cars will they need? _____

4. Denzel has $15.00. He wants to buy as many baseballs as he can. If each baseball costs $4.00, how many can he buy, and how much money will he have left over? _____

5. April has a book case with 5 shelves. Each shelf will hold 6 dolls from her collection of 28 dolls. How many shelves will be full, and how many dolls will go on the last shelf? ____

Divide and show your remainders in your answers. (DOK 2)

6. $4\overline{)61}$ 10. $7\overline{)814}$ 14. $8\overline{)49}$

7. $3\overline{)67}$ 11. $2\overline{)87}$ 15. $9\overline{)815}$

8. $6\overline{)127}$ 12. $4\overline{)91}$ 16. $3\overline{)632}$

9. $5\overline{)128}$ 13. $3\overline{)23}$ 17. $2\overline{)51}$

Divide and show your remainders in your answers. (DOK 2)

18. $826 \div 13$ 19. $473 \div 22$ 20. $638 \div 27$ 21. $951 \div 8$

_____   _____   _____   _____

2.4 Relationship Between Multiplication and Division (DOK 2)

There is a relationship between multiplication and division. The two factors in multiplication yield a product. The product divided by either of the factors equals the other factor.

Example 7: $3 \times 25 = 75$ 3 and 25 are the factors, 75 is the product.

$75 \div 3 = 25$ and $75 \div 25 = 3$

The product from the first problem divided by either factor equals the other factor.

Find the two division number sentences that can be made from the factors and product of the multiplication problems. (DOK 2)

1. $33 \times 4 = 132$
2. $20 \times 5 = 100$
3. $15 \times 6 = 90$
4. $5 \times 9 = 45$
5. $11 \times 9 = 99$
6. $150 \times 5 = 750$
7. $14 \times 6 = 84$
8. $22 \times 3 = 66$
9. $8 \times 7 = 56$
10. $9 \times 3 = 27$
11. $19 \times 22 = 418$
12. $91 \times 2 = 182$
13. $5 \times 11 = 55$
14. $28 \times 16 = 448$
15. $40 \times 4 = 160$

Dividend ÷ Divisor = Quotient   Divisor) Dividend (Quotient)

Example 8: $90 \div 2 = 45$

$2 \times 45 = 90$ and $45 \times 2 = 90$

Find the two multiplication number sentences that can be made from each division problem. (DOK 2)

16. $16 \div 8 = 2$
17. $48 \div 12 = 4$
18. $99 \div 9 = 11$
19. $140 \div 5 = 28$
20. $36 \div 4 = 9$
21. $44 \div 2 = 22$
22. $28 \div 7 = 4$
23. $120 \div 10 = 12$
24. $540 \div 90 = 6$
25. $72 \div 8 = 9$
26. $15 \div 5 = 3$
27. $8 \div 4 = 2$
28. $65 \div 5 = 13$
29. $24 \div 8 = 3$
30. $100 \div 5 = 20$

Chapter 2   Whole Number Division

2.5   Modeling Division (DOK 2)

$15 \div 5 = 3$

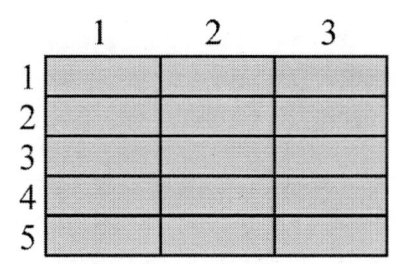

The model above shows 1 rectangle divided into 15 parts.
The 15 parts are arranged into 3 columns of 5 rows. $3 \times 5 = 15$ $15 \div 3 = 5$
Or, the 15 parts are arranged into 5 rows of 3 columns. $5 \times 3 = 15$ $15 \div 5 = 3$

Write two multiplication problems and two division problems and solve for each model below. (DOK 2)

1.

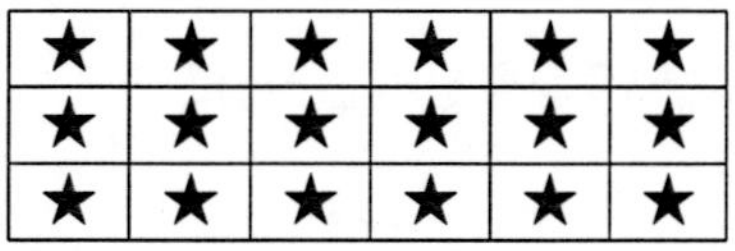

2.

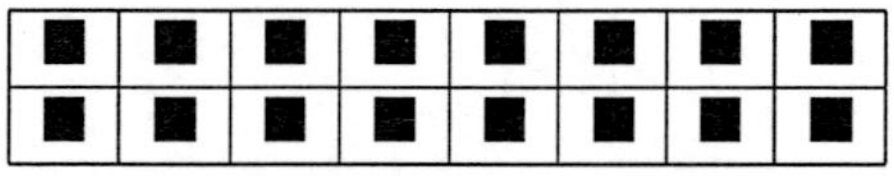

3.

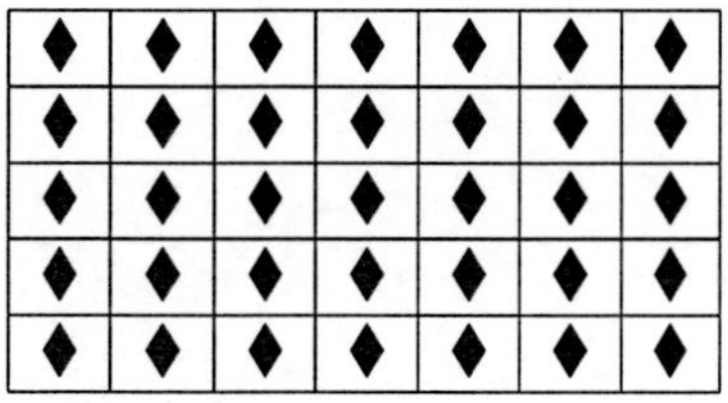

4.

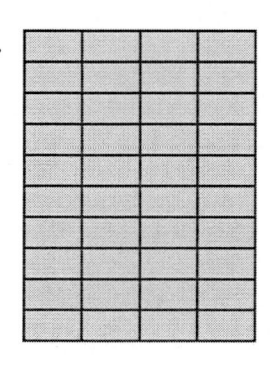

5.

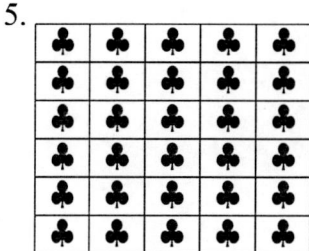

6.

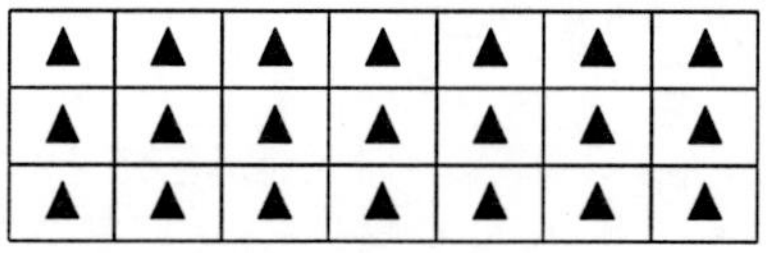

7.

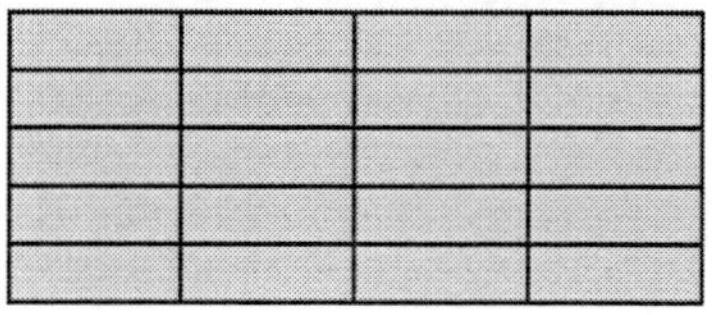

8.

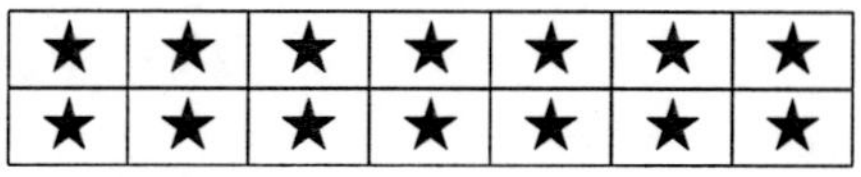

2.6 Going Deeper into Whole Number Division (DOK 3)

Solve the multi-step word problems below. Show your work for each step. (DOK 3)

1. A farmer estimates that he has 1,200,000 pounds of watermelon ready to harvest. The average weight of his watermelons is 12 pounds.

   Part 1: Estimate the number of watermelons the farmer will harvest.

   Part 2: If the farmer delivers an equal number of watermelons to the warehouses of 4 grocery chain stores, how many watermelons will each grocery chain receive?

2. The fifth grade teachers at Sunny Elementary are given one case of paper to share. There are 4 teachers and one case has 10 reams of 500 sheets of paper.

   Part 1: How many sheets of paper will each teacher receive?

   Part 2: If each teacher has 24 students, how many sheets of paper will be used for each student and how many sheets will be left over (remainder)?

3. The chart below shows the number of different kinds of book printed and bound by the Bountiful Book Bindery in one week.

Kind of Book	Number
Arts & Crafts	1,100
Cookbooks	1,785
Gardening	955
Mechanical	850

   Part 1: If the number of books bound for the week are evenly divided into 5 crates to send to 5 stores, how many of each kind of book will go to each store?

   Part 2: What is the total number of books that will go to each store?

4. Mr. Blake is a gym teacher. He is holding tryouts for the boys basketball team. Starting with dribbling relays, he takes out 12 basketballs for the 58 boys to show their abilities. If he divides the boys into 12 groups, one basketball for each group, how many boys will be in each group? (Hint: Not all the groups will have the same number of players.)

5. A company shows profits of $78,020 for the first 3 months of the year. The company has 47 employees and the owner of the company has decided that one quarter of the profit will be evenly divided among the employees as a bonus. How much money will each employee receive?

6. Eileen bakes 360 cookies to sell at the school fair. She puts 3 cookies in each plastic baggy. The school fair lasts 6 hours. She sells the same number of bags of cookies each hour of the fair.

   Part 1: How many bags of cookies did Eileen sell each hour?

   Part 2: How many cookies did she sell the first four hours?

7. Noah makes 240 blueberry muffins for the same school fair. He sells an equal number of muffins in each of the 6 hours the school fair lasts.

   Part 1: How many blueberry muffins did Noah sell each hour?

   Part 2: How many blueberry muffins did Noah sell the first 3 hours?

Chapter 2   Whole Number Division

Chapter 2 Review

Review the numbers below. Write whether they are divisible by 2, 3, 5, or 10. You may have more than one answer. (DOK 1)

1. 30 _____
2. 48 _____
3. 52 _____
4. 66 _____

Divide. (DOK 2)

5. $355 \div 71$ _____
6. $784 \div 14$ _____
7. $576 \div 32$ _____
8. $240 \div 15$ _____
9. $870 \div 30$ _____
10. $3,430 \div 70$ _____
11. $1,881 \div 33$ _____
12. $1,722 \div 82$ _____
13. $5,032 \div 74$ _____

Divide. Be certain to include the remainders in your answers. (DOK 2)

14. $741 \div 12$ _____
15. $683 \div 34$ _____
16. $851 \div 26$ _____
17. $229 \div 42$ _____
18. $347 \div 13$ _____
19. $26 \div 8$ _____

Find the <u>two</u> division number sentences that can be made from the factors and product of the multiplication problems. (DOK 2)

20. $42 \times 4 = 168$
21. $21 \times 7 = 147$
22. $13 \times 6 = 78$

Find the <u>two</u> multiplication number sentences that can be made from each division problem. (DOK 2)

23. $35 \div 7 = 5$
24. $54 \div 6 = 9$
25. $100 \div 5 = 20$

Write <u>two</u> division problems and solve for each model below. (DOK 2)

26.

27.

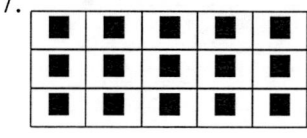

Solve the multi-step division problem below. Show your work. (DOK 3)

28. A company manufactures 4,750 skateboards each day. On Wednesday, the entire 4,750 skateboards were sent in equal amounts to 5 different department store chains.

   Part 1: How many skate boards went to each department store chain?

   Part 2: If one of the department store chains divided their share of the skateboards between their stores in 5 different states, how many skateboards would they send to each state?

Chapter 2 Test

Choose the correct answer.

1 Divide: 216 ÷ 72

  A 30
  B 3
  C 13
  D 23

(DOK 2)

2 Divide: 400 ÷ 16

  A 30
  B 25
  C 35
  D 23

(DOK 2)

3 Divide: 588 ÷ 42

  A 17
  B 15
  C 12
  D 14

(DOK 2)

4 Divide: 783 ÷ 13

  A 62 R3
  B 60 R1
  C 60 R3
  D 61 R3

(DOK 2)

5 Divide: 354 ÷ 12

  A 29 R6
  B 29 R8
  C 28 R6
  D 28 R8

(DOK 2)

6 Divide: 112 ÷ 7

  A 16
  B 16 R1
  C 16 R2
  D 17 R1

(DOK 2)

7 Divide: 652 ÷ 24

  A 27 R3
  B 26 R2
  C 27 R2
  D 27 R4

(DOK 2)

8 Which sentence is true?

  A A number is divisible by 2 if it ends in a zero or an even number.

  B A number is divisible by 3 if it ends in a zero or an even number.

  C A number is always divisible by 5 if it ends in 3, 6, or 9.

  D A number is always divisible by 3 if it ends in 1, 5, or 7.

(DOK 1)

9 Find the <u>two</u> division number sentences that can be made from the factors and product of the multiplication problem:

$3 \times 5 = 15$.

  A $15 \div 5 = 3$, $15 \times 3 = 45$
  B $5 \times 3 = 15$, $15 \div 3 = 5$
  C $15 \div 5 = 3$, $15 \div 3 = 5$
  D $15 \div 5 = 3$, $3 \times 5 = 15$

(DOK 2)

Chapter 2   Whole Number Division

10 Find the two multiplication number sentences that can be made from the division problem:

$48 \div 8 = 6$.

A $6 \times 8 = 48, 8 \times 6 = 48$

B $6 \times 8 = 48, 8 \times 8 = 48$

C $6 \times 6 = 36, 8 \times 8 = 64$

D $48 \div 6 = 8, 8 \times 6 = 48$

(DOK 2)

11 Divide: $3,018 \div 12$

A 251 R6

B 252 R6

C 251 R9

D 252 R7

(DOK 2)

12 What division sentence is represented by the model below?

A $8 \times 10 = 80$

B $80 \div 8 = 8$

C $80 \div 10 = 8$

D $80 \div 10 = 10$

(DOK 2)

13 A school cafeteria is preparing lunch that will include baby carrots for 750 of the students at Larson Elementary. The remainder of the students brought their own lunch. A case of baby carrots holds 50 pounds of baby carrots. Each pound has 30 baby carrots. The school prepares 3 cases of baby carrots. How many baby carrots will each of the 750 students buying a school lunch receive?

A 3

B 15

C 12

D 6

(DOK 3)

14 Which number is divisible evenly by 2 and 3?

A 895

B 894

C 896

D 897

(DOK 2)

15 What division sentence is represented by the model below?

A $6 \times 5 = 30$

B $30 \div 6 = 5$

C $6 + 6 + 6 + 6 + 6 = 30$

D $30 - 6 - 6 - 6 - 6 - 6 = 0$

(DOK 2)

Chapter 3
Decimals

This chapter covers the following CC 5 standards:

| Number and Operations in Base 10 | 5.NBT.3, 5.NBT.4 |

3.1 Place Value: Less Than One (DOK 1, 2)

Place Value Chart from 1 to 0.001

Ones		Tenths	Hundredths	Thousandths
1.0		$\frac{1}{10}$ or 0.1	$\frac{1}{100}$ or 0.01	$\frac{1}{1,000}$ or 0.001
1	.	9	8	7

Let's review the place value chart from left to right. The ones column was discussed earlier. Now, let's see what happens when we look at the numbers after the decimal point.

From the left, let us begin with the tenths column. As we move from column to column, we divide by 10. Example 0.1 divided by 10 = 0.01

The number above in standard form: 1.987

Expanded Decimal Form: $1 + 0.9 + 0.08 + 0.007$

Chapter 3   Decimals

Example 1:   Review the chart on the previous page. Which number is in the hundredths place?

Answer: 8 is in the hundredths place. The value is calculated as follows:

$8 \times 0.01 = 0.08$

Example 2:   Which number is in the tenths place?

Answer: 9 is in the tenths place. The value is calculated as follows:

$9 \times 0.1 = 0.9$

What is the place value of the number 2 in the following decimal numbers? (DOK 1)

1. 0.028 _____
2. 0.214 _____
3. 0.002 _____
4. 0.126 _____
5. 0.182 _____
6. 0.253 _____
7. 1.912 _____
8. 0.02 _____
9. 1.2 _____

Review the numbers below and write the place value of the underlined digit. (DOK 1)

10. 32.89<u>3</u> _____
11. 4.0<u>7</u>6 _____
12. 63.86<u>7</u> _____
13. 1.00<u>4</u> _____
14. 75.<u>6</u>18 _____
15. 39.88<u>6</u> _____

Use the numbers below to answer the questions that follow. (DOK 2)

| 2,116.750 | 9,824.375 | 3,572.143 | 6,721.854 | 7,223.008 |

16. In which number is the 7 in the hundredths column? _____

17. In which number is the 7 in the tenths column? _____

18. In which number is the 8 in the thousandths column? _____

19. In which number is the 4 in the hundredths column? _____

20. In which number is the 8 in the tenths column? _____

3.2 Reading and Writing Decimal Numbers (DOK 1)

Example 3: Write the number 162.748 in words.

Hundreds	Tens	Ones	.	Tenths	Hundredths	Thousandths
1	6	2	.	7	4	8

Refer to the "About AR" on page xi!

Step 1: Write the number to the left of the decimal point:
One hundred sixty-two.

Step 2: Write the word "and" where the decimal point is:
One hundred sixty-two and.

Step 3: Write the number after the decimal point and add the name of the column of the last digit:
One hundred sixty-two and seven hundred forty-eight thousandths.

Answer: One hundred sixty-two and seven hundred forty-eight thousandths.

Note: Only use the word "and" where the decimal point comes. NEVER write one hundred and sixty-two for the number 162. Just write one hundred sixty-two.

Write the following numbers in words. (DOK 1)

1. 882.434 _____

2. 999.9 _____

3. 41.756 _____

4. 722.31 _____

5. 161.227 _____

Use digits to write each of the following whole numbers and decimals. Remember the word "and" means put in a decimal point. (DOK 1)

6. _____ sixty-seven and fifteen hundredths

7. _____ five hundred twenty-two and four tenths

8. _____ seven hundred seventy-one and seven hundred seventy-one thousandths

9. _____ eighty-four and ninety-seven hundredths

10. _____ three hundred fifty-five and eight hundred forty-six thousandths

Chapter 3 Decimals

3.3 Expanded Form of Decimals (DOK 2)

Decimal numbers can be separated into parts using multiplication. This is called the **expanded form of decimals**.

Example 4: Expand the decimal number 487.253.

Step 1: Multiply each of the whole number digits by its place value.
$(4 \times \underline{100}) + (8 \times \underline{10}) + (7 \times \underline{1})$

Step 2: Multiply each of the place values that are less than 1.
$(2 \times \frac{1}{10}) + (5 \times \frac{1}{100}) + (3 \times \frac{1}{1000})$

Answer: $487.253 = (4 \times \underline{100}) + (8 \times \underline{10}) + (7 \times \underline{1}) + (2 \times \frac{1}{10}) + (5 \times \frac{1}{100}) + (3 \times \frac{1}{1000})$

A zero in a place value does not have to be multiplied.

Expand the decimal numbers below using multiplication. Write your answers on your own paper. (DOK 2)

1. 22.17
2. 349.553
3. 61.48
4. 10.007
5. 9.183
6. 12.121
7. 8,007.701
8. 952.33
9. 677.1
10. 4.828
11. 0.135
12. 520.025
13. 62.773
14. 44.004
15. 79.911
16. 804.048
17. 35.503
18. 6.719
19. 0.002
20. 100.305

Write the decimal number from the following multiplication problems. (DOK 2)

21. $(8 \times 100) + (2 \times 1) + (7 \times \frac{1}{10}) + (3 \times \frac{1}{100}) + (1 \times \frac{1}{1,000})$ _____

22. $(6 \times 100) + (4 \times \frac{1}{10}) + (2 \times \frac{1}{100}) + (7 \times \frac{1}{1,000})$ _____

23. $(9 \times 1,000) + (2 \times 100) + (3 \times 10) + (6 \times \frac{1}{10}) + (5 \times \frac{1}{100}) + (9 \times \frac{1}{1,000})$ _____

24. $(2 \times 100) + (3 \times 10) + (7 \times 1) + (2 \times \frac{1}{10}) + (7 \times \frac{1}{1,000})$ _____

25. $(1 \times 1,000) + (2 \times 100) + (7 \times 10) + (2 \times 1) + (3 \times \frac{1}{10}) + (1 \times \frac{1}{100}) + (8 \times \frac{1}{1,000})$ _____

3.4 Estimating Decimals Using Rounding (DOK 2)

Example 5: Estimate 8.5 to the nearest **whole number**.

Step 1: Look at the number in the tenths place. If the number in the tenths place is 5 or greater, then the number in the ones column will round up to the next whole number. If the number in the tenths place is less than 5, the number will round down and the whole number in the ones column will remain the same.

Step 2: Because the number in the tenths place is 5, we round the whole number from 8 to 9.

Answer: 9

Example 6: Estimate 12.87 to the nearest **tenth**.

Step 1: Look at the hundredths place. Since 7 is greater than 5, round the tenths place up.

Step 2: The tenths place increases by one from 8 to 9.

Answer: 12.9

Round the decimals to the nearest whole number. (DOK 2)

1. 11.4 _____
2. 278.9 _____
3. 56.31 _____
4. 12.515 _____
5. 1,458.7 _____
6. 56.3 _____
7. 78.6 _____
8. 32.04 _____
9. 5.72 _____
10. 16.51 _____

Round the decimals to the nearest tenth. (DOK 2)

11. 62.14 _____
12. 8.77 _____
13. 6.35 _____
14. 10.08 _____
15. 9.23 _____
16. 21.92 _____
17. 34.08 _____
18. 17.71 _____
19. 233.06 _____
20. 5.55 _____

Round the decimals to the nearest hundredth. (DOK 2)

21. 7.625 _____
22. 12.333 _____
23. 6.267 _____
24. 3.241 _____
25. 8.928 _____
26. 5.555 _____
27. 43.017 _____
28. 30.002 _____
29. 91.191 _____
30. 51.066 _____

Chapter 3 Decimals

3.5 Ordering Decimals (DOK 2)

Example 7: Order the following decimals from greatest to least.

0.3, 0.208, 0.34

Step 1: Arrange numbers with decimal points directly under each other.

0.3
0.208
0.34

Step 2: Fill in with zeros so they all have the same number of places after the decimal point. Remember to **read the numbers as if the decimal points were not there.**

0.300
0.208 ← Least
0.340 ← Greatest

Answer: 0.34, 0.3, 0.208

Order each set of decimals below from greatest to least. (DOK 2)

1. 0.075   0.705   0.7   0.75
2. 0.5   0.56   0.65   0.06
3. 0.9   0.09   0.099   0.95

4. 0.6   0.59   0.06   0.66
5. 0.3   0.303   0.03   0.33
6. 0.02   0.25   0.205   0.5

Order each set of decimals below from least to greatest. (DOK 2)

7. 0.055   0.5   0.59   0.05
8. 0.7   0.732   0.74   0.72
9. 0.04   0.48   0.048   0.408

10. 0.9   0.905   0.95   0.09
11. 0.19   0.09   0.9   0.1
12. 0.21   0.02   0.021   0.2

At a math competition, teams won by giving the correct answer to a math problem in the shortest amount of time. The chart below gives the names and times of each team for the first problem.

Matherines	1.625 minutes
Numberamas	1.605 minutes
Math-maniacs	1.615 minutes
Number-jumbles	1.675 minutes

13. Which team took the longest time to solve the problem? _____

14. Which team took the shortest time to solve the problem? _____

15. Which team came in second place for this problem? _____

16. Which team came in third place for this problem? _____

3.6 Comparing Decimals (DOK 2)

Example 8: Compare 4.1 and 4.11.

Step 1: Arrange numbers with decimal points directly under each other.

4.1
4.11

Step 2: Fill in with zeros so they all have the same number of digits after the decimal point. **Read the numbers as if the decimal points were not there.**

4.10
4.11

Answer: 4.1 < 4.11

Fill in the box with the correct sign (>, <, or =). (DOK 2)

1. 7.2 ☐ 7.02
2. 0.30 ☐ 0.3
3. 9.44 ☐ 9.4
4. 21.01 ☐ 21.11
5. 6.22 ☐ 6.20
6. 17.1001 ☐ 17.1011
7. 8.55 ☐ 8.555
8. 29.06 ☐ 29.060
9. 40.99 ☐ 40.9
10. 97.08 ☐ 97.18
11. 0.21 ☐ 0.211
12. 36.66 ☐ 36.606

Use the list of numbers below to answer the questions that follow. (DOK 2)

| 7.50 | 1.00 | 14.530 | 0.990 | 0.049 | 15.44 |

13. Which two numbers for R makes this statement true? $R \geq 14.53$

14. Which number for P makes this statement true? $P < 0.056$

15. Which number for N makes this statement true? $N = 7.5$

16. Which number for B makes this statement true? $B > 15.43$

17. Which three numbers for S makes this statement true? $S \leq 1.01$

18. Which number for W makes this statement true? $W = 0.99$

Chapter 3   Decimals

3.7   Going Deeper into Decimals (DOK 3)

Read each multi-step problem and solve. Show your work for each step. (DOK 3)

1. The table below shows the weights at two months of age, of 4 puppies belonging to Abby, the Labrador.

Puppy	Weight in Pounds
Chewy	3.47
Louie	4.25
Betty	3.95
Lottie	4.13

   Part 1: Estimate the average weight of each puppy to the nearest whole pound.

   Part 2: Put the weights of the four puppies in order from smallest to largest. Just use the name of the puppy in your answer.

2. Charlie weighs $(8 \times 10) + (4 \times 1) + (7 \times \frac{1}{10}) + (4 \times \frac{1}{100})$ pounds. His older brother, Harley, weighs $(1 \times 100) + (2 \times 10) + (1 \times 1) + (4 \times \frac{1}{10}) + (8 \times \frac{1}{100})$ pounds.

   Part 1: Give the weights of both boys in standard form.

   Part 2: Round both weights to the nearest tenth of a pound.

3. Trisha's cat, Meowly, is 5.625 years old. Her dog, Barker, is 5.55 years old.

   Part 1: Give both weights in expanded form and compare the two weights using a greater than or less than sign.

   Part 2: Round both weights to the nearest tenth of a pound.

4. Erin gathered the seashells she found on the beach each day of vacation and put them in a bag. She weighed the bags and put the data in the table below.

Day	Weight of Seashells
Monday	11.375 lb
Tuesday	9.65 lb
Wednesday	10.75 lb
Thursday	7.5 lb

   Part 1: Round the weights of the seashells for each day to the nearest whole pound. Then find the total weight of the seashells for the four days, using the rounded weights.

   Part 2: Give the weight of the seashells on Monday in expanded form.

5. Dereck wrote to his grandfather to let him know he caught a fish that weighed 7.852 pounds.

   Part 1: Write the weight of the fish in expanded form.

   Part 2: Round the weight of the fish to the nearest tenth of a pound.

6. There are $(9 \times 1,000) + (3 \times 100) + (4 \times \frac{1}{100})$ dollars left in Mrs. Scofield's savings account.

   Part 1: Write the amount in standard form.

   Part 2: Round the amount to the nearest thousand dollars.

Chapter 3 Review

Write the place value of the digit 7 in the following decimal fractions. (DOK 1)

1. 298.071 _____
2. 15.72 _____
3. 14.357 _____

4. Write the word form of 824.76. _____

5. Write the word form of 32.044. _____

6. Write the number form of: twelve and three hundred sixty-seven thousandths. _____

7. Write the number form of seven and thirty-two hundredths. _____

Order each set of decimals below from least to greatest. (DOK 2)

8. 4.84 4.75 4.08 4.48
9. 0.33 3.33 0.03 0.3

Order each set of decimals below from greatest to least. (DOK 2)

10. 6.72 0.23 6.23 0.27
11. 15.11 11.15 51.11 11.51

Write the decimal number from the multiplication problems below. (DOK 2)

12. $(9 \times 1,000) + (4 \times 10) + (3 \times \frac{1}{10}) + (3 \times \frac{1}{100}) + (2 \times \frac{1}{1,000})$ _____

13. $(2 \times 100) + (5 \times 10) + (1 \times 1) + (2 \times \frac{1}{10}) + (7 \times \frac{1}{1,000})$ _____

14. $(6 \times 1,000) + (2 \times 100) + (1 \times \frac{1}{10}) + (8 \times \frac{1}{100}) + (4 \times \frac{1}{1,000})$ _____

Fill in the box with the correct sign (>, <, or =). (DOK 2)

15. 6.3 ☐ 6.031
16. 0.89 ☐ 0.904
17. 4.011 ☐ 40.1
18. 34.23 ☐ 34.023

Expand the decimal numbers below using multiplication. (DOK 2)

19. 3,200.824 _____

20. 590.055 _____

Round the numbers below. (DOK 2)

21. Round to nearest tenth: 87.09 _____
22. Round to nearest hundredth: 15.284 _____

Chapter 3   Decimals

Chapter 3 Test

1 What place value is the 8 in 961.85?

A hundred
B hundredths
C tens
D tenths

(DOK 1)

2 Put the following decimal fractions in order from **least** to **greatest**.

   4.52   4.4   4.24   4.3

A 4.52   4.3   4.4   4.24
B 4.24   4.3   4.4   4.52
C 4.3   4.4   4.24   4.52
D 4.24   4.52   4.3   4.4

(DOK 2)

3 Which number is equal to:

$(2 \times 10) + (3 \times \frac{1}{10}) + (4 \times \frac{1}{1,000})$?

A 200.304
B 20.304
C 20.34
D 200.34

(DOK 2)

4 Which is the expanded form of 809.02?

A $(8 \times 100) + (9 \times 1) + (2 \times \frac{1}{100})$

B $(8 \times 10) + (9 \times 1) + (2 \times \frac{1}{100})$

C $(8 \times 100) + (9 \times 10) + (2 \times \frac{1}{100})$

D $(8 \times 100) + (9 \times 1) + (2 \times \frac{1}{10})$

(DOK 2)

5 Round 844.361 to the nearest whole number.

A 842
B 843.4
C 844
D 844.4

(DOK 2)

6 What place value is the 2 in 645.102?

A ones
B tenths
C hundredths
D thousandths

(DOK 1)

7 What is eight thousand thirteen and two hundred fifty-four thousandths in number form?

A 8,013.245
B 8,310.254
C 8,030.245
D 8,013.254

(DOK 1)

8 What is 704.126 in word form?

A seven hundred four and one hundred sixty-two thousandths
B seven hundred four and one hundred twenty-six thousandths
C seven hundred forty and one hundred twenty-six thousandths
D seven hundred forty and one hundred sixty-two thousandths

(DOK 1)

9 Round 78.57 to the nearest tenth.

A 78.5
B 78.6
C 79.6
D 78.7

(DOK 1)

Chapter 3 Test

10 Which of the following is in order from **least** to **greatest**?

A	3.13	3.03	3.01
B	4.04	4.14	4.44
C	5.15	5.05	5.01
D	6.12	6.22	6.02

(DOK 2)

11 Which of the following is in order from **greatest** to **least**?

A	8.13	8.01	8.03
B	9.09	9.19	9.49
C	4.14	4.04	4.01
D	7.13	7.23	7.03

(DOK 2)

12 Which of the following comparisons is true?

- **A** $0.4 < 0.04$
- **B** $0.04 < 0.4$
- **C** $0.4 > 0.44$
- **D** $0.04 > 0.4$

(DOK 2)

13 Which of the following comparisons is true?

- **A** $1.15 > 1.05$
- **B** $1.05 > 1.15$
- **C** $1.51 < 1.5$
- **D** $1.5 > 1.51$

(DOK 2)

14 What is the place value of 3 in 21.834?

- **A** tens
- **B** tenths
- **C** hundredths
- **D** thousandths

(DOK 1)

15 Which number is equal to:

$(7 \times 100) + (2 \times \frac{1}{10}) + (5 \times \frac{1}{100})$?

- **A** 700.25
- **B** 70.25
- **C** 700.025
- **D** 70.025

(DOK 2)

16 Which of the following is in order from **least** to **greatest**?

A	6.25	6.05	6.15
B	7.25	7.35	7.55
C	8.1	8.11	8.01
D	5.05	5.55	5.5

(DOK 2)

17 Which of the following comparisons is true?

- **A** $6.02 > 6.01$
- **B** $6.12 > 6.22$
- **C** $6.22 < 6.21$
- **D** $6.22 > 6.23$

(DOK 2)

18 Which of the following is in order from **greatest** to **least**?

A	6.76	6.66	6.77
B	7.77	7.87	7.97
C	8.28	8.88	8.22
D	9.99	9.91	9.19

(DOK 2)

19 Maria weighs $(7 \times 10) + (9 \times 1) + (6 \times \frac{1}{10}) + (5 \times \frac{1}{100})$. Her brother, Enrique, weighs $(1 \times 100) + (2 \times 10) + (8 \times 1) + (9 \times \frac{1}{10})$. Which sentence below correctly identifies both weights in standard form and correctly compares the two numbers?

- **A** $79.65 > 128.9$
- **B** $79.56 > 128.09$
- **C** $79.65 < 128.9$
- **D** $79.56 > 128.09$

(DOK 3)

Chapter 4
Performing Operations on Decimals

This chapter covers the following CC 5 standards:

| Number and Operations in Base 10 | 5.NBT.2, 5.NBT.7 |

4.1 Adding Decimals (DOK 1)

Example 1: Add: $7.8 + 1.16$

Step 1: When adding decimals, first arrange the numbers in columns with the decimal points directly under each other. If necessary, fill in with zeros to the right of the number, so all the numbers have the same amount of places after the decimal point.

Step 2: Start at the right and add each column. Remember to carry when necessary. Bring the decimal point straight down in the answer. Add $7.8 + 1.16$.

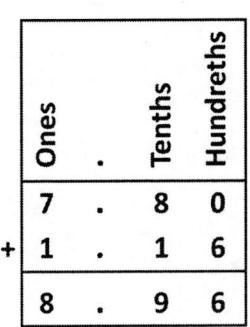

Ones	.	Tenths	Hundredths
7	.	8	0
+ 1	.	1	6
8	.	9	6

Answer: 8.96

Add the following. Be sure to write the decimal point in your answer. (DOK 1)

1. $8.24 + 6.3$ _____
2. $5.4 + 0.81$ _____
3. $4.27 + 0.22$ _____
4. $2.99 + 4.99$ _____
5. $1.17 + 7.17$ _____
6. $6.32 + 0.24$ _____
7. $0.73 + 0.52$ _____
8. $4.58 + 0.24$ _____
9. $2.32 + 0.64$ _____
10. $8.68 + 0.08$ _____
11. $0.6 + 0.06$ _____
12. $7.45 + 0.2$ _____

4.2 Subtracting Decimals (DOK 1)

Example 2: Subtract: 9.45 − 6.2

Step 1: When you subtract decimals, first arrange the numbers in columns with the decimal points under each other. If necessary, fill in with zeros to the right of the number, so all the numbers have the same amount of places after the decimal point.

Step 2: Subtract as you would any subtraction problem. Remember to borrow when needed. Bring the decimal point straight down in the answer. Subtract 9.45 − 6.2.

Ones	.	Tenths	Hundreths
9	.	4	5
− 6	.	2	0
3	.	2	5

Answer: 3.25

Subtract the following. Be sure to write the decimal in your answer. (DOK 1)

1. 12.1 − 3.04 _____
2. 9.65 − 4.21 _____
3. 0.22 − 0.21 _____
4. 0.89 − 0.05 _____
5. 7.72 − 2.27 _____
6. 11.54 − 9.65 _____
7. 34.05 − 2.5 _____
8. 4.6 − 3.02 _____
9. 54.54 − 45.45 _____
10. 10.1 − 0.11 _____
11. 86.32 − 27.42 _____
12. 12.04 − 1.09 _____

Chapter 4   Performing Operations on Decimals

4.3   Relationship Between Addition and Subtraction of Decimals (DOK 2)

There is a relationship called a **fact family** that is a group of three numbers which are related through addition and subtraction.

Example 3:   Find the fact family for 3.2, 1.4, and 4.6.

Addition number sentences:
$$3.2 + 1.4 = 4.6$$
$$1.4 + 3.2 = 4.6$$

Subtraction number sentences:
$$4.6 - 3.2 = 1.4$$
$$4.6 - 1.4 = 3.2$$

Find the four number sentences that make up the fact families for the following decimal numbers. (DOK 2)

1. 5.2, 6.8, 12
2. 14.1, 5.2, 19.3
3. 2.5, 3.7, 6.2
4. 10.8, 5.2, 5.6
5. 18.2, 6.3, 24.5
6. 3.7, 12.5, 8.8
7. 7.5, 8.1, 15.6
8. 4.2, 2.4, 6.6
9. 19.8, 11.2, 31
10. 14.2, 7.5, 6.7
11. 3.3, 2.2, 5.5
12. 108.7, 51.6, 57.1
13. 11.1, 33.3, 22.2
14. 9.015, 12.5, 3.485
15. 1.26, 4.322, 5.582
16. 1.111, 6.771, 5.66
17. 2.85, 3.42, 6.27
18. 9.002, 17.172, 8.17
19. 4.4, 8.5, 4.1
20. 12.64, 6.21, 6.43
21. 0.005, 0.05, 0.055
22. 8.416, 3.274, 11.69
23. 0.11, 1.001, 1.111
24. 5.5, 2.73, 8.23

4.4 Multiplying Decimals by Multiples of Ten (DOK 2)

In chapter one, you learned to multiply whole numbers by multiples of ten. For example, $54 \times 10^3 = 54,000$. The exponent, in this case 3, tells you how many zeros to add to the other factor, 54.

The same can be done with decimal numbers. Take the exponent from the power of ten and add those zeros to the decimal number factor. Then move the decimal point to the right the same number of places as the exponent.

Example 4: $4.5 \times 10^2 = ?$

Step 1: 10^2 has two zeros. We are going to move the decimal place over two places to the right.

4.50

Step 2: When you move the decimal place over two places to the right, you move once to the right after the five and then one more time and add one zero. 4.5 becomes 450.
Answer: 450

Example 5: $880 \times 10^3 = ?$

Step 1: 10^3 has 3 zeros. We are going to move the decimal place over three places to the right.

880.000

Refer to the "About AR" on page xi!

Step 2: When you move the decimal place over three to the right, you must put three zeros after 880. You get 880,000.
Answer: 880,000

Find the product. (DOK 2)

1. 27.8×10^4 _____
2. 1.35×10^2 _____
3. 18.76×10^3 _____
4. 50.7×10^5 _____
5. 86.5×10^2 _____
6. 4.44×10^3 _____
7. 9.87×10^5 _____
8. 14.92×10^2 _____
9. 1.6×10^6 _____
10. 0.68×10^4 _____
11. 3.21×10^2 _____
12. 6.458×10^5 _____
13. 0.001×10^2 _____
14. 7.04×10^4 _____
15. 11.11×10^6 _____

Chapter 4 Performing Operations on Decimals

4.5 Multiplying Decimals by Whole Numbers (DOK 2)

Example 6: 87×2.4

Step 1:   If written horizontally, rewrite the problem vertically.

$$\begin{array}{r} 87 \\ \times 2.4 \\ \hline \end{array}$$

Step 2:   Multiply as if all numbers are whole. Disregard the decimal point, for now.

$$\begin{array}{r} 87 \\ \times 2.4 \\ \hline 348 \\ +1740 \\ \hline 2088 \end{array}$$

Step 3:   For each factor, count the numbers that appear after the decimal point.

$$\begin{array}{ll} 87 & \text{0 numbers after the decimal} \\ 2.4 & +\text{ 1 number after the decimal} \\ & \overline{\text{1 number after the decimal in total}} \end{array}$$

Step 4:   If there is one number after the decimal point in the problem, there should be one number after the decimal point in the answer.
Product should have one (0 + 1) number after the decimal: 208.8

Answer: 208.8

Multiply. (DOK 2)

1. 6.7×8 _____
2. 18.9×7 _____
3. 21.1×4 _____
4. $1,122 \times 1.2$ _____
5. 85.8×9 _____
6. 48×3.41 _____
7. 12×7.9 _____
8. 54×66.51 _____
9. 95×14.2 _____
10. 111×7.31 _____
11. 21×14.112 _____
12. 13×6.72 _____
13. 45×10.431 _____
14. 62×2.87 _____
15. 99×1.111 _____
16. 7×103.42 _____
17. 18×2.301 _____
18. 84×3.3 _____
19. 55×1.55 _____
20. 17×2.82 _____
21. 53×6.12 _____
22. 101×7.82 _____
23. 41×3.7 _____
24. 186×1.45 _____

4.6 Multiplying Decimals by Decimals (DOK 2)

Example 7: 56.2×0.17

Step 1: Set up the problem as if you were multiplying whole numbers.

$$\begin{array}{r} 56.2 \\ \times\, 0.17 \\ \hline \end{array}$$

Step 2: Multiply as if you were multiplying whole numbers.

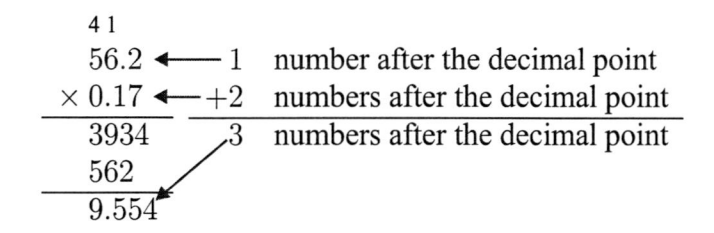

```
    4 1
   56.2   ←— 1 number after the decimal point
 × 0.17   ←—+2 numbers after the decimal point
  ─────
   3934    3 numbers after the decimal point
   562
  ─────
   9.554
```

Step 3: Count how many numbers are after the decimal points in the problem. In this problem, three numbers, 2, 1, and 7, come after decimal points, so the answer must also have three numbers after the decimal point.
Answer: 9.554

Multiply. (DOK 2)

1. 15.2×3.5 _____
2. 9.54×5.3 _____
3. 5.72×6.3 _____
4. 4.8×3.2 _____
5. 45.8×2.2 _____
6. 4.5×7.1 _____
7. 0.52×0.3 _____
8. 4.12×6.8 _____

9. 23.65×9.2 _____
10. 1.54×0.4 _____
11. 0.47×6.1 _____
12. 1.3×1.57 _____
13. 16.4×0.5 _____
14. 0.87×3.2 _____
15. 5.94×0.65 _____
16. 7.8×0.23 _____

17. 4.1×5.23 _____
18. 1.23×0.1 _____
19. 9.5×3.25 _____
20. 1.2×0.87 _____
21. 9.91×1.6 _____
22. 3.3×0.2 _____
23. 7.5×8.21 _____
24. 12.4×6.35 _____

Chapter 4   Performing Operations on Decimals

4.7   More Multiplying Decimals by Decimals (DOK 2)

Example 8:   Find 0.007×0.125

Step 1: Multiply as you would whole numbers.

$$\begin{array}{r} 0.007 \\ \times\, 0.125 \\ \hline 0.000875 \end{array}$$

← 3 numbers after the decimal point
← +3 numbers after the decimal point
← 6 numbers after the decimal point

Step 2: Count how many numbers are to the right of the decimal points in the problem. In this case, six numbers come after decimal points in the problem, so there must be six numbers after the decimal point in the answer. In this problem, 0's needed to be written in the answer in front of the 8, so there would be 6 numbers after the decimal point.
Answer: 0.000875

Multiply. Write in zeros as needed. Round dollar figures to the nearest penny. (DOK 2)

1. 0.123×0.45 _____

2. 0.004×10.31 _____

3. 1.54×1.1 _____

4. 10.05×0.45 _____

5. 9.45×0.8 _____

6. $\$6.49 \times 0.06$ _____

7. 5.003×0.009 _____

8. $\$9.99 \times 0.06$ _____

9. 6.09×5.3 _____

10. $\$22.00 \times 0.075$ _____

11. 5.914×0.02 _____

12. 4.96×0.23 _____

13. 6.98×0.02 _____

14. 3.12×0.08 _____

15. 7.158×0.09 _____

16. 0.0158×0.32 _____

4.8 Division of Decimals by Whole Numbers (DOK 2)

Example 9: $52.26 \div 6$

Step 1: Copy the problem as you would for whole numbers. Copy the decimal point directly above in the place for the answer.

$$6 \overline{) 52.26}$$

Step 2: Divide the same way as you would with whole numbers.

```
      8 . 7 1
6 ) 5 2 . 2 6
    4 8
      4   2
    − 4   2
            6
          − 6
            0
```

Divide. Remember to copy the decimal point directly above the place for the answer.

1. $42.75 \div 3$ _____
2. $74.16 \div 6$ _____
3. $81.50 \div 25$ _____
4. $82.46 \div 14$ _____
5. $12.50 \div 2$ _____
6. $224.64 \div 52$ _____
7. $183.04 \div 52$ _____
8. $281.52 \div 23$ _____
9. $72.36 \div 4$ _____
10. $379.5 \div 15$ _____
11. $152.25 \div 21$ _____
12. $40.375 \div 19$ _____
13. $102.5 \div 5$ _____
14. $113.4 \div 9$ _____
15. $585.14 \div 34$ _____
16. $93.6 \div 24$ _____

Chapter 4 Performing Operations on Decimals

4.9 Division of Decimals by Decimals (DOK 2)

Example 10: $374.5 \div 0.07$

Step 1: Copy the problem as you would for whole numbers.

$0.07 \overline{) 374.5}$ ← Dividend (Divisor points to 0.07)

Step 2: You cannot divide by a decimal number. You must move the decimal point in the divisor 2 places to the right to make it a whole number. The decimal point in the dividend must also move to the right the same number of places. Notice that in this example, you must add a 0 to the dividend.

$0.07. \overline{) 374.50.}$

Step 3: The problem now becomes $37,450 \div 7$. Copy the decimal point from the dividend straight above in the place for the answer.

```
         5 3 5 0.
      7 )3 7 4 5 0.
       - 3 5
           2 4
         - 2 1
             3 5
           - 3 5
               0 0
```

Divide. Remember to move the decimal points. (DOK 2)

1. $0.676 \div 0.013$ _____
2. $70.32 \div 0.08$ _____
3. $54.60 \div 0.84$ _____
4. $10.35 \div 0.45$ _____
5. $18.46 \div 1.3$ _____
6. $14.6 \div 0.002$ _____
7. $125.25 \div 0.75$ _____
8. $33.00 \div 1.65$ _____
9. $154.08 \div 1.8$ _____
10. $0.4374 \div 0.003$ _____
11. $292.9 \div 0.29$ _____
12. $6.375 \div 0.3$ _____
13. $4.8 \div 0.08$ _____
14. $1.2 \div 0.024$ _____
15. $15.725 \div 3.7$ _____
16. $167.50 \div 0.25$ _____

4.10 Modeling Operations on Decimals (DOK 2)

The model below represents the number 1 divided into 100 parts. There are 100 squares within the one block. Each square within the model below represents 0.01.

100 squares × 0.01 = 1.

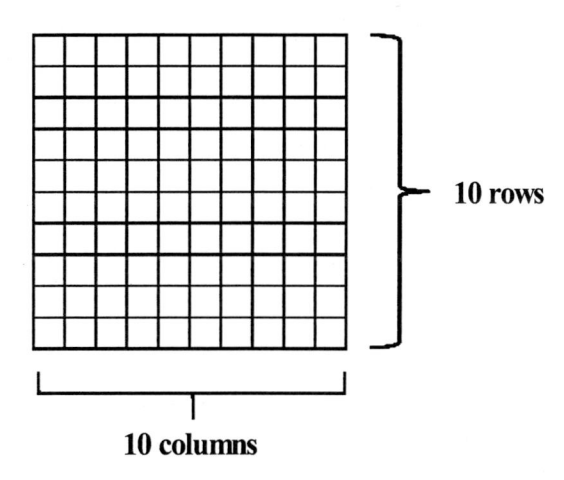

Each shaded portion of the blocks represents a decimal.

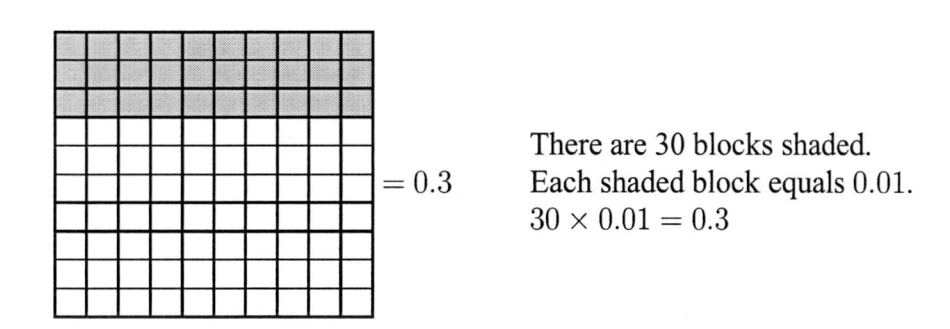

= 0.3   There are 30 blocks shaded.
Each shaded block equals 0.01.
30 × 0.01 = 0.3

Solve the problems below using the shaded portion of the models. There are addition, subtraction, multiplication, and division problems. (DOK 2)

1.   +   _____

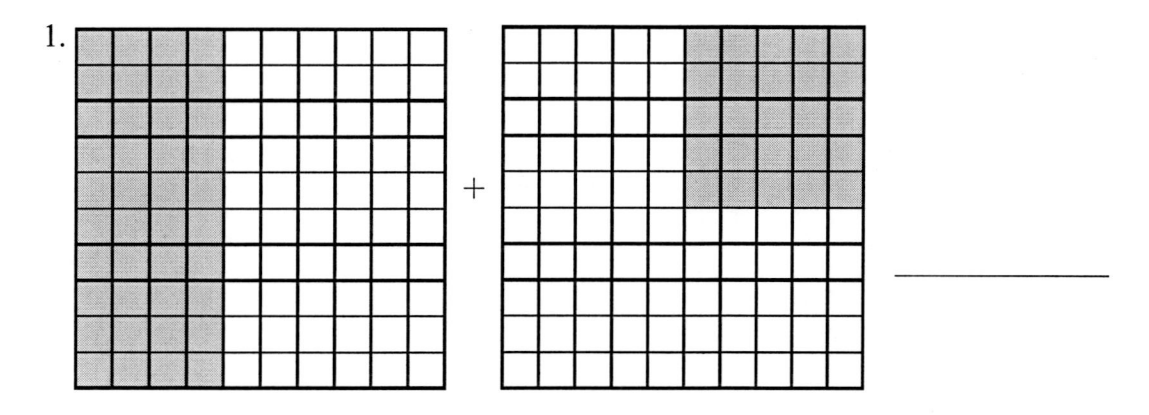

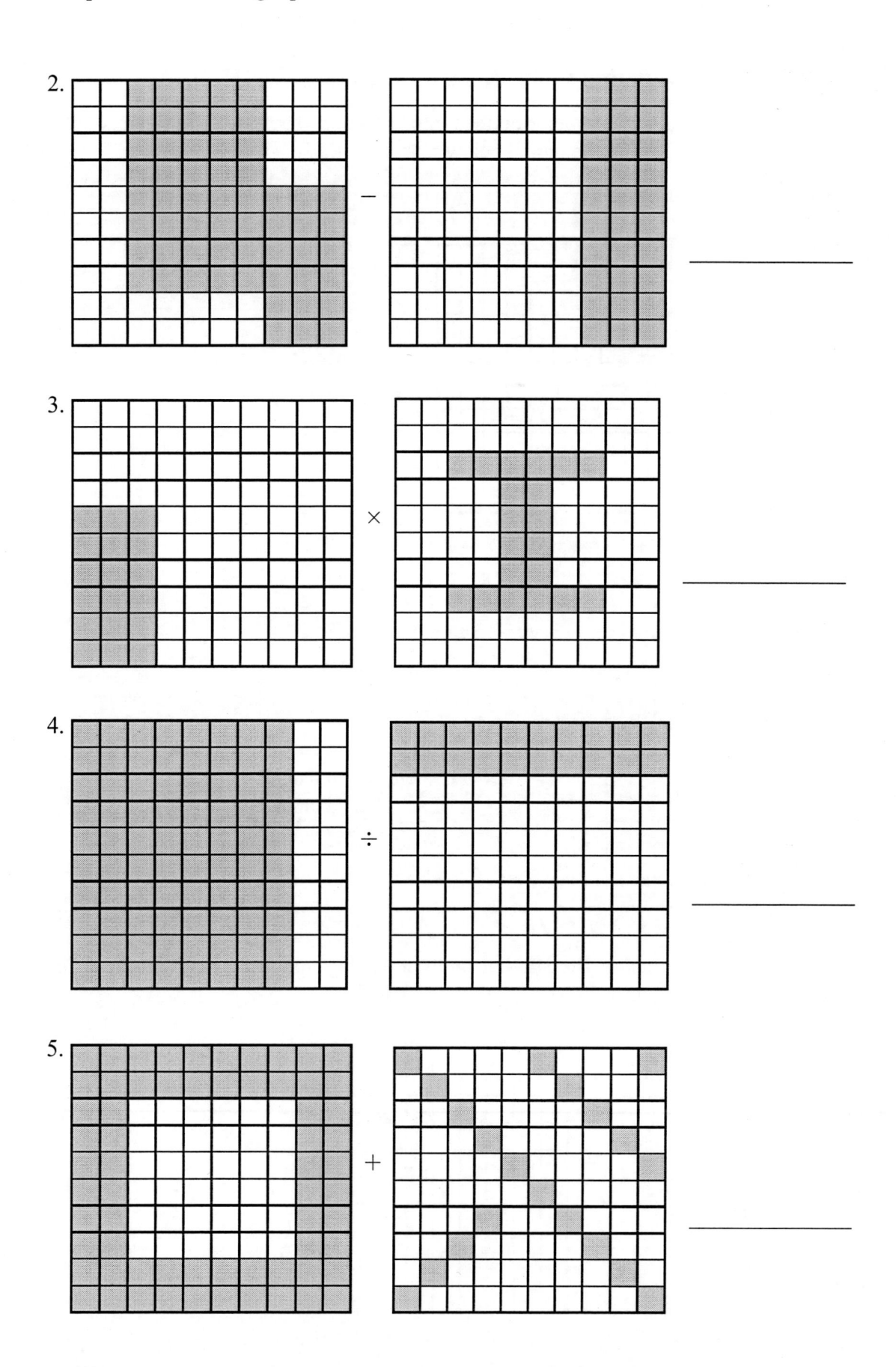

Chapter 4 Performing Operations on Decimals

2. [grid] − [grid] = _____

3. [grid] × [grid] = _____

4. [grid] ÷ [grid] = _____

5. [grid] + [grid] = _____

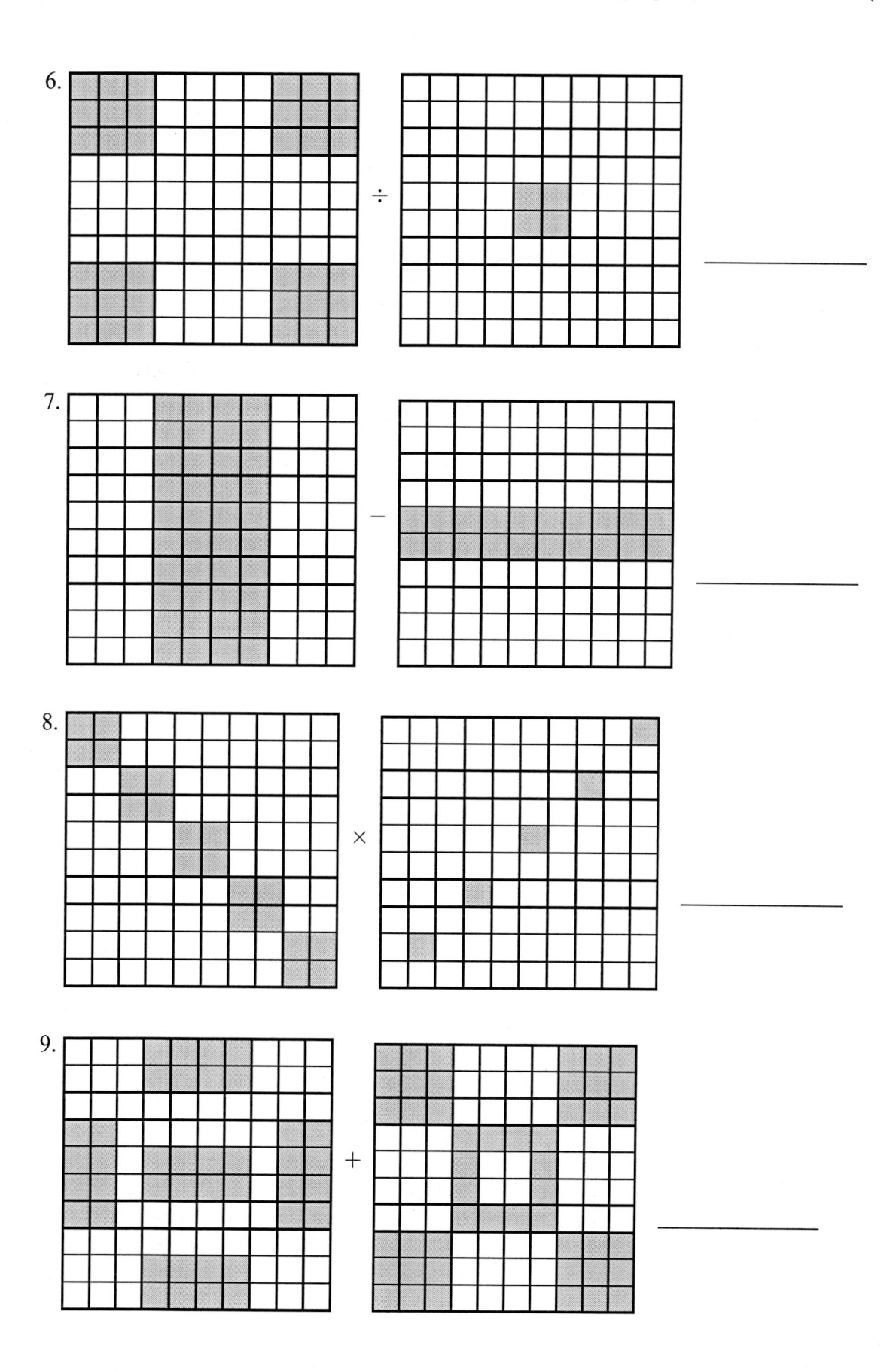

4.10 Modeling Operations on Decimals (DOK 2)

6. ⬚ ÷ ⬚ = _____

7. ⬚ − ⬚ = _____

8. ⬚ × ⬚ = _____

9. ⬚ + ⬚ = _____

Chapter 4 Performing Operations on Decimals

4.11 Going Deeper into Performing Operations on Decimals (DOK 3)

Solve the multi-step problems below. Show your work for each step. (DOK 3)

1. Jonas and Silas are brothers that started a car wash business in the driveway of their parents home for the summer. Jonas washes the cars and Silas rinses and dries the cars. The boys decided to divide the money at the end of each day according to the model below.

   Key:

Jonas	Shaded grids
Silas	White grids

   Part 1: At the end of day one, the boys earned a total of $40.00. How much money did each boy earn?

   Part 2: At the end of the first week, the boys earned $280.00. How much money did each boy earn?

2. A bag of apples at the grocery store costs $3.99, before taxes. The grocery store sells 17 bags of apples on Friday.

   Part 1: How much money did the grocery store take in for the 17 bags of apples?

   Part 2: The grocery store paid $3.25, before taxes, for each bag of apples. How much did the grocery store make on the apples to pay towards expenses such as salaries, rent, utilities, etc.?

3. A case of video games cost the department store $5,478.00, including taxes. There are 50 video games in each case.

   Part 1: How much does each video game cost the department store?

   Part 2: If the mark up on the video games to the customer is $23.87, to cover expenses of the store and for profit, how much will the customers pay for all 50 video games?

4. Lisa is buying a blouse that costs $24.99 and a pair of jeans that cost $16.97. She gives the cashier $40.00.

   Part 1: Does Lisa have enough money to pay for the blouse and jeans? Find the amount she is short or the amount she will receive as change.

   Part 2: If Lisa receives $3.75 each week for an allowance, how many weeks did it take her to earn the $40.00? Give your answer to the nearest tenth.

5. Jalisha bought a box of 20 packages of beads at a yard sale for $5.15. She makes 8 necklaces from the beads and yarn she has leftover from a scarf she knitted. Jalisha sells the necklaces for $3.00 each.

   Part 1: How much money did Jalisha receive for the 8 necklaces?

   Part 2: What was the profit Jalisha made on the necklaces? The profit is found by taking the money Jalisha received from the sale of the necklaces minus the amount she paid for the beads.

Chapter 4 Review

Solve the decimal problems below. (DOK 1)

1. $4.32 + 7.125$ _____
2. $8.77 - 2.5$ _____
3. $12.895 + 11.235$ _____
4. $9.842 - 6.15$ _____
5. $3.27 + 4.822$ _____
6. $18.74 - 13.006$ _____

Find the four number sentences of fact families for the decimal numbers below. (DOK 2)

7. 15.5, 7.2, 8.3
8. 2.35, 14.7, 12.35
9. 8.711, 5.301, 3.41

Find the product. (DOK 2)

10. 11.41×10^4 _____
11. 2.01×10^2 _____
12. 33.303×10^5 _____

Multiply. (DOK 2)

13. 4.7×3 _____
14. 24.512×6 _____
15. 78.436×2 _____

Multiply. (DOK 2)

16. 41.6×2.1 _____
17. 1.315×8.2 _____
18. 6.2×3.21 _____

Divide. (DOK 2)

19. $15.9 \div 3$ _____
20. $54.12 \div 6$ _____
21. $175.75 \div 25$ _____

Divide. (DOK 2)

22. $2.385 \div 0.05$ _____
23. $99.9 \div 0.03$ _____
24. $34.52 \div 0.2$ _____

Solve the problems below using the shaded portion of the models. (DOK 2)

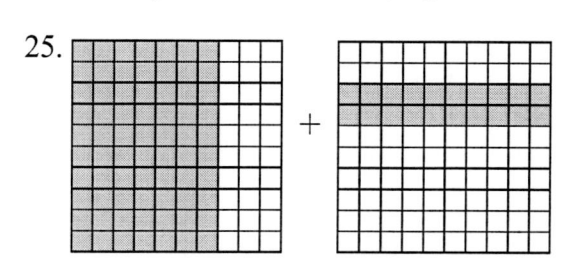

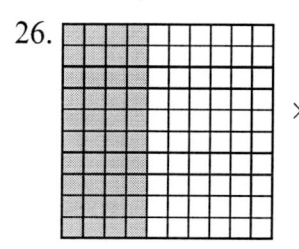

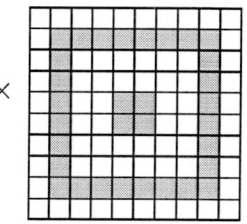

25.

26.

Solve the multi-step problem below. Show your work. (DOK 3)

27. Katie bought 4 different yarns for $2.79 each, including tax. She made 8 scarves and sold them to her friends for $6.00 each. How much profit did Katie make on the 4 scarves? Profit equals the money she received from the sale of the scarves minus the amount of money she paid for the yarn.

Chapter 4 Performing Operations on Decimals

Chapter 4 Test

1 Add: $75.632 + 14.215$

  A 88.847
  B 88.947
  C 89.847
  D 61.417

(DOK 1)

2 Subtract: $112.556 - 83.42$

  A 29.136
  B 195.976
  C 30.136
  D 28.135

(DOK 1)

3 Multiply: $487 \times 2.2 =$

  A 107.41
  B 107.14
  C 1,071.4
  D 1,071.14

(DOK 2)

4 Multiply: $6.27 \times 0.4 =$

  A 2.668
  B 2.558
  C 2.608
  D 2.508

(DOK 2)

5 Divide:

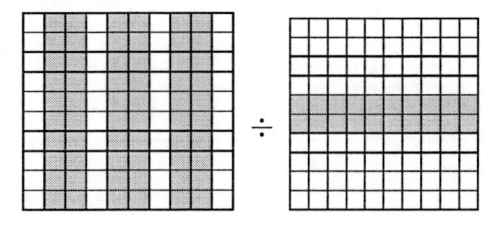

  A 0.003
  B 0.03
  C 0.3
  D 3

(DOK 2)

6 Multiply: 5.42×10^3

  A 542
  B 5,420
  C 54,200
  D 542,000

(DOK 2)

7 Divide: $124 \div 6.2 =$

  A 20.12
  B 20
  C 20.22
  D 20.02

(DOK 2)

8 Divide: $16.8 \div 2.1 =$

  A 8.0
  B 8.8
  C 8.4
  D 8.2

(DOK 2)

9 Add:

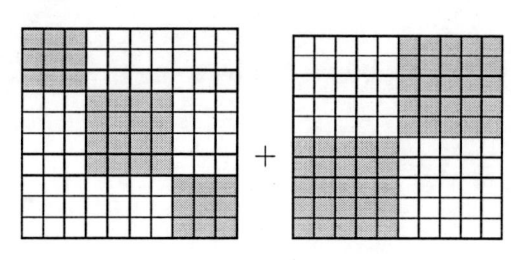

  A 0.68
  B 1.00
  C 0.84
  D 0.85

(DOK 2)

Chapter 4 Test

10 What is the missing member of the fact family:

| 7.5 + 18.2 = 25.7 |
| 18.2 + 7.5 = 25.7 |
| 25.7 − 7.5 = 18.2 |

A $7.5 \times 18.2 = 25.7$
B $25.7 - 18.2 = 7.5$
C $25.7 - 7.5 = 18.2$
D $25.7 \div 7.5 = 18.2$

(DOK 2)

11 What is the missing member of the fact family:

| 34.813 − 17.54 = 17.273 |
| 34.813 − 17.273 = 17.54 |
| 17.54 + 17.273 = 34.813 |

A $17.273 + 17.54 = 34.813$
B $34.813 \div 17.54 = 17.273$
C $17.54 \times 17.273 = 34.813$
D $34.813 - 17.54 = 17.273$

(DOK 2)

12 Add: $9.007 + 1.003$

A 100.01
B 100.1
C 10.01
D 10.001

(DOK 1)

13 Multiply: 2.258×10^5

A 0.00002258
B 2,258
C 22,580
D 225,800

(DOK 2)

14 Multiply: 4.47×10^4

A 4,470
B 44,700
C 447,000
D 4,470,000

(DOK 2)

15 Subtract: $8.799 - 4.311$

A 4.884
B 4.388
C 4.488
D 13.11

(DOK 1)

16 Divide: $8.972 \div 0.02$

A 0.4486
B 4.486
C 44.86
D 448.6

(DOK 2)

17 Multiply: 6.24×5

A 11.24
B 31.2
C 31.24
D 31.95

(DOK 2)

18. Kenneth bought $18.57 of supplies to make birdhouses to sell to neighbors. He sold all the birdhouses he made according to the table below.

Size of Birdhouse	Number Sold	Amount
Small	3	$4.25
Medium	4	$6.50
Large	2	$8.25

How much profit did Kenneth make? Profit is the money received for sales minus the cost of materials.

A $33.68
B $46.68
C $55.25
D $36.68

(DOK 3)

Chapter 5
Adding and Subtracting Fractions

This chapter covers the following CC 5 standards:

| Number and Operations - Fractions | 5.NF.1, 5.NF.2 |

5.1 Fractions (DOK 1)

Fractions are numbers used to express part of a total.

Fractions are expressed as part(s) over the total. $\frac{\text{Part}}{\text{Total}}$

Example 1: Sara's mother cuts a pie into 5 equal pieces. How much of the pie is one slice?

Step 1: First determine the number of parts. The part is 1 piece of pie. The total is 5 pieces of pie.

Answer: $\frac{\text{Part}}{\text{Total}} = \frac{1}{5}$ One slice of the pie is $\frac{1}{5}$ of the pie.

The names of the two parts of a fraction are **numerator** for the part on top and **denominator** for the total on the bottom.

$$\frac{\text{Numerator}}{\text{Denominator}} \quad \frac{\text{Part}}{\text{Total}} \quad \frac{1}{5}$$

An easy way to remember which names of numerator and denominator go where: denominator starts with the letter "d" and so does the word "down." The denominator is down from the numerator.

If the numerators are all equal, and the denominators are different in a group of fractions, then the larger the denominator, the smaller the fraction. Look at the list of fractions below.

$$\frac{1}{2} \quad \frac{1}{3} \quad \frac{1}{4} \quad \frac{1}{5} \quad \frac{1}{6} \quad \frac{1}{8}$$

Think of a pie. If the pie is cut into two pieces, each piece is $\frac{1}{2}$ of a pie. That's a big piece of pie!

If the pie is cut into eight pieces, each piece is $\frac{1}{8}$ of a pie. That's a much smaller piece of pie than $\frac{1}{2}$ of a pie.

5.2 Simplifying Fractions (DOK 1)

Example 2: Reduce $\frac{4}{8}$ to lowest terms.

Step 1: First you need to find the greatest common factor of 4 and 8. Think: What is the largest number that can be divided into 4 and 8 without a remainder?

These must be the same number. $?\overline{)4}$ $?\overline{)8}$ 4 and 8 can both be divided by 4.

Step 2: Divide the top (numerator) and bottom (denominator) of the fraction by the same number.

$$\frac{4 \div 4}{8 \div 4} = \frac{1}{2} \quad \text{Therefore, } \frac{4}{8} = \frac{1}{2}.$$

If there is a whole number with the fraction, be certain to carry the whole number forward with the simplified fraction.

Examples: $3\frac{3}{6}$ simplified is $3\frac{1}{2}$ $15\frac{7}{21}$ simplified is $15\frac{1}{3}$ $347\frac{9}{15}$ simplified is $347\frac{3}{5}$

Simplify the following fractions. (DOK 1)

1. $\frac{2}{8}$ _____
2. $\frac{12}{15}$ _____
3. $\frac{9}{27}$ _____
4. $\frac{12}{42}$ _____
5. $\frac{3}{21}$ _____
6. $\frac{27}{54}$ _____
7. $\frac{14}{22}$ _____

8. $\frac{9}{21}$ _____
9. $\frac{4}{14}$ _____
10. $\frac{6}{26}$ _____
11. $\frac{30}{45}$ _____
12. $\frac{16}{64}$ _____
13. $\frac{10}{25}$ _____
14. $\frac{3}{12}$ _____

15. $\frac{15}{30}$ _____
16. $\frac{12}{36}$ _____
17. $\frac{13}{39}$ _____
18. $\frac{28}{49}$ _____
19. $\frac{8}{18}$ _____
20. $\frac{14}{21}$ _____
21. $1\frac{2}{12}$ _____

22. $3\frac{5}{15}$ _____
23. $4\frac{9}{15}$ _____
24. $7\frac{24}{48}$ _____
25. $8\frac{3}{18}$ _____
26. $9\frac{6}{27}$ _____
27. $15\frac{4}{18}$ _____
28. $8\frac{8}{28}$ _____

Chapter 5 Adding and Subtracting Fractions

5.3 Simplifying Improper Fractions (DOK 1, 2)

Example 3: Simplify $\frac{21}{4}$.

Step 1: $\frac{21}{4} = 21 \div 4 = 5$ remainder 1

The quotient, 5, becomes the whole number portion of the mixed number.

Step 2: Make the remainder, 1, become the top number of the fraction.

$\frac{21}{4} = 5\frac{1}{4}$ ← remainder

The denominator (bottom number) of the fraction always remains the same.

Example 4: Simplify $\frac{11}{6}$.

Step 1: $\frac{11}{6}$ is the same as $11 \div 6$. $11 \div 6 = 1$ with a remainder of 5.

Step 2: Rewrite as a mixed number. $1\frac{5}{6}$

Simplify the following improper fractions. (DOK 1)

1. $\frac{13}{5} = $ _____

2. $\frac{11}{3} = $ _____

3. $\frac{24}{6} = $ _____

4. $\frac{7}{6} = $ _____

5. $\frac{19}{6} = $ _____

6. $\frac{16}{5} = $ _____

7. $\frac{13}{8} = $ _____

8. $\frac{9}{5} = $ _____

9. $\frac{22}{3} = $ _____

10. $\frac{13}{4} = $ _____

11. $\frac{15}{2} = $ _____

12. $\frac{22}{6} = $ _____

13. $\frac{17}{8} = $ _____

14. $\frac{27}{8} = $ _____

15. $\frac{32}{5} = $ _____

16. $\frac{3}{2} = $ _____

17. $\frac{7}{4} = $ _____

18. $\frac{21}{10} = $ _____

19. $\frac{12}{2} = $ _____

20. $\frac{17}{3} = $ _____

21. $\frac{23}{10} = $ _____

22. $\frac{9}{2} = $ _____

23. $\frac{12}{5} = $ _____

24. $\frac{19}{3} = $ _____

5.3 Simplifying Improper Fractions (DOK 1, 2)

Now take what you learned simplifying improper fractions a step further, by adding whole numbers to the fractions.

Example 5: Simplify $8\frac{3}{2}$.

Step 1: Separate the whole number from the improper fraction. $8\frac{3}{2}$ separates to 8 and $\frac{3}{2}$.

Step 2: Simplify the improper fraction $\frac{3}{2}$. $\frac{3}{2}$ is the same as $3 \div 2$. $3 \div 2 = 1$ with a remainder of 1.

Step 3: Rewrite $\frac{3}{2}$ as a whole number with a fraction. $1\frac{1}{2}$.

Step 4: Add $1\frac{1}{2}$ to the whole number 8 from the original problem. $8 + 1\frac{1}{2} = 9\frac{1}{2}$.

Simplify the following improper mixed numbers. (DOK 2)

1. $6\frac{12}{8} =$ _____

2. $11\frac{7}{5} =$ _____

3. $9\frac{10}{5} =$ _____

4. $2\frac{3}{2} =$ _____

5. $5\frac{11}{6} =$ _____

6. $8\frac{15}{6} =$ _____

7. $10\frac{5}{2} =$ _____

8. $7\frac{7}{3} =$ _____

9. $11\frac{11}{4} =$ _____

10. $2\frac{9}{5} =$ _____

11. $3\frac{13}{3} =$ _____

12. $6\frac{4}{3} =$ _____

13. $22\frac{11}{2} =$ _____

14. $1\frac{5}{4} =$ _____

15. $3\frac{8}{5} =$ _____

16. $5\frac{17}{3} =$ _____

17. $15\frac{5}{2} =$ _____

18. $9\frac{6}{5} =$ _____

19. $1\frac{11}{3} =$ _____

20. $12\frac{7}{3} =$ _____

21. $9\frac{4}{2} =$ _____

Chapter 5 Adding and Subtracting Fractions

5.4 Equivalent Fractions (DOK 1, 2)

Remember, any fraction that has the same non-zero numerator (top number) and denominator (bottom number) equals 1.

Example 6: $\frac{5}{5} = 1 \quad \frac{8}{8} = 1 \quad \frac{12}{12} = 1 \quad \frac{15}{15} = 1 \quad \frac{25}{25} = 1$

Any fraction multiplied by 1 in any form remains equal to itself.

Example 7: $\frac{3}{7} \times \frac{4}{4} = \frac{12}{28} = \frac{3}{7}$

Example 8: Find the missing numerator (top number). $\frac{5}{8} = \frac{}{24}$

Step 1: Ask yourself, "What was 8 multiplied by to get 24?" 3 is the answer.

Step 2: The only way to keep the fraction equal is to multiply the top and bottom numbers by the same number. The bottom number was multiplied by 3, so multiply the top number by 3, as shown below.

$\frac{5}{8} \times \frac{3}{3} = \frac{15}{24}$ Note: $\frac{3}{3} = 1$

Find the missing numerators from the following equivalent fractions. (DOK 1)

1. $\frac{2}{6} = \frac{}{18}$

2. $\frac{2}{3} = \frac{}{27}$

3. $\frac{4}{9} = \frac{}{18}$

4. $\frac{7}{15} = \frac{}{45}$

5. $\frac{9}{10} = \frac{}{50}$

6. $\frac{5}{6} = \frac{}{36}$

7. $\frac{1}{4} = \frac{}{36}$

8. $\frac{3}{14} = \frac{}{28}$

9. $\frac{2}{5} = \frac{}{25}$

10. $\frac{4}{11} = \frac{}{33}$

11. $\frac{5}{6} = \frac{}{18}$

12. $\frac{6}{11} = \frac{}{22}$

13. $\frac{8}{15} = \frac{}{45}$

14. $\frac{1}{9} = \frac{}{18}$

15. $\frac{7}{8} = \frac{}{40}$

16. $\frac{1}{12} = \frac{}{48}$

17. $\frac{3}{8} = \frac{}{24}$

18. $\frac{3}{4} = \frac{}{16}$

19. $\frac{2}{7} = \frac{}{49}$

20. $\frac{11}{12} = \frac{}{24}$

21. $1\frac{2}{5} = 1\frac{}{45}$

22. $2\frac{4}{5} = 2\frac{}{15}$

23. $8\frac{1}{9} = 8\frac{}{27}$

24. $3\frac{3}{8} = 3\frac{}{56}$

25. $11\frac{2}{13} = 11\frac{}{26}$

26. $7\frac{1}{7} = 7\frac{}{35}$

27. $2\frac{4}{5} = 2\frac{}{10}$

28. $1\frac{3}{10} = 1\frac{}{40}$

5.4 Equivalent Fractions (DOK 1, 2)

Choose the answers from the chart below to answer the questions that follow. (DOK 2)

A) $\frac{1}{2}$	B) $\frac{3}{8}$	C) $\frac{4}{5}$	D) $\frac{9}{10}$	E) $\frac{19}{20}$	F) $\frac{31}{50}$	G) $\frac{43}{80}$	H) $\frac{111}{250}$	I) $\frac{357}{1,000}$	J) $\frac{990}{10,000}$

1. Marcus cut a pan of brownies into 16 pieces. He served 6 pieces for he and his two friends to share. Which fraction above is equal to the fraction of the pan of brownies Marcus served?

2. Mr. Tillman works in a factory and packs glass ornaments into cartons, each carton holding 1,000 ornaments. So far, he has packed 444 ornaments into the box he is currently packing. Which fraction above is equal to the number of ornaments in the box so far?

3. Melanie got 95 out of 100 questions on the math test correct. Which fraction above equals the fraction of questions Melanie answered correctly?

4. Jason is practicing his jump shot while shooting basketballs into the net. Out of 160 tries, he makes 86 baskets. Which fraction above is equal to the number of baskets he made?

5. Isabella and Samantha share a package of gum containing 14 sticks, equally. Which fraction above shows the amount of gum Samantha gets?

6. For every 1,000 people at the stadium, 99 have on red shirts. Which fraction above shows the number of red shirts worn by people in a crowd of 10,000?

7. There is a common saying in commercials that "9 out of 10 doctors recommend..." some product. Which fraction above shows the amount of doctors who may recommend a certain product?

8. The Sassy Sock Company shipped 3,000 pairs of socks to a large department store. Of those, 1,071 pairs had a black stripe at the top of the sock. Which fraction above is equal to the number of pairs of socks shipped with a black stripe at the top?

9. Andrew has a goal to earn $100.00 over summer break. It's mid-July, and he has earned $62.00 so far. Which fraction above is equal to the amount of money he has earned towards his goal so far?

10. There are 20 students on the bus. Sixteen of the students are in 4th or 5th grade. Which fraction above is equal to the number of students on the bus who are in either 4th or 5th grade?

Copyright © American Book Company

Chapter 5   Adding and Subtracting Fractions

5.5   Adding Fractions (DOK 2)

When **adding fractions** that have the same denominator, the denominator stays the same in the answer. You simply add the numerators and simplify the resulting sum if necessary.

Example 9:   Add: $3\frac{4}{5} + 2\frac{2}{5}$

Step 1:   Rewrite the problem vertically, and add down.
$$3\frac{4}{5}$$
$$+2\frac{2}{5}$$

Step 2:   Add the whole numbers, $3 + 2 = 5$. And add the fractions, $\frac{4}{5} + \frac{2}{5} = \frac{6}{5}$.
$$3\frac{4}{5}$$
$$+2\frac{2}{5}$$
$$5\frac{6}{5}$$

Step 3:   Simplify $\frac{6}{5}$ to $1\frac{1}{5}$, and add 1 to the whole number, 5, in the answer.

Step 4: $5\frac{6}{5} = 5 + 1\frac{1}{5} = 6\frac{1}{5}$

Add and simplify the answers. (DOK 2)

1. $3\frac{5}{6}$
    $+5\frac{2}{6}$

3. $3\frac{3}{5}$
    $+2\frac{3}{5}$

5. $6\frac{5}{6}$
    $+4\frac{1}{6}$

7. $\frac{1}{4}$
    $+7\frac{3}{4}$

9. $4\frac{7}{10}$
    $+8\frac{2}{10}$

11. $3\frac{3}{8}$
    $+2\frac{3}{8}$

2. $1\frac{1}{4}$
    $+4\frac{2}{4}$

4. $2\frac{1}{8}$
    $+1\frac{7}{8}$

6. $9\frac{1}{4}$
    $+5\frac{5}{4}$

8. $9\frac{4}{8}$
    $+3\frac{2}{8}$

10. $5\frac{2}{5}$
    $+\frac{1}{5}$

12. $\frac{3}{6}$
    $+\frac{4}{6}$

5.6 Subtracting Fractions (DOK 2)

When **subtracting fractions** that have the same denominator, the denominator stays the same in the answer. You simply subtract the numerators and simplify the resulting difference, if necessary.

Example 10: Subtract: $9\frac{7}{8} - 2\frac{1}{8}$

Step 1: Rewrite the problem vertically, then subtract.

$$\begin{array}{r} 9\frac{7}{8} \\ -2\frac{1}{8} \\ \hline \end{array}$$

Step 2: Subtract the fractions, $\frac{7}{8} - \frac{1}{8} = \frac{6}{8}$. And subtract the whole numbers, $9 - 2 = 7$.

$$\begin{array}{r} 9\frac{7}{8} \\ -2\frac{1}{8} \\ \hline 7\frac{6}{8} \end{array}$$

Step 3: Simplify $\frac{6}{8}$ to $\frac{3}{4}$

Step 4: $9\frac{7}{8} - 2\frac{1}{8} = 7\frac{6}{8} = 7\frac{3}{4}$

Subtract and simplify the answers. (DOK 2)

1. $4\frac{7}{8}$
   $-1\frac{3}{8}$

2. $6\frac{9}{10}$
   $-\frac{2}{10}$

3. $\frac{7}{8}$
   $-\frac{4}{8}$

4. $13\frac{3}{4}$
   $-7\frac{2}{4}$

5. $14\frac{9}{10}$
   $-8\frac{1}{10}$

6. $3\frac{5}{6}$
   $-\frac{1}{6}$

7. $5\frac{4}{5}$
   $-3\frac{2}{5}$

8. $\frac{5}{6}$
   $-\frac{1}{6}$

5.7 Least Common Multiple (DOK 2)

Find the least common multiple (LCM) of 6 and 10.

To find the least common multiple (LCM) of two numbers, first list the multiples of each number. The multiples of a number are 1 times the number, 2 times the number, 3 times the number, etc.

The multiples of 6 are: 6, 12, 18, 24, 30...

The multiples of 10 are: 10, 20, 30, 40, 50...

What is the smallest multiple they both have in common?   30.

30 is the least (smallest number) common multiple of 6 and 10.

Find the least common multiple (LCM) of each pair of numbers below. (DOK 2)

	Pairs	Multiples	LCM		Pairs	Multiples	LCM
1.	6	6, 12, 18, 24, 30	30	10.	6		
	15	15, 30			7		
2.	12			11.	7		
	16				14		
3.	18			12.	9		
	36				5		
4.	7			13.	30		
	3				45		
5.	12			14.	3		
	8				8		
6.	6			15.	12		
	8				9		
7.	4			16.	5		
	14				35		
8.	9			17.	3		
	6				5		
9.	2			18.	4		
	7				12		

5.8 Adding Mixed Numbers (DOK 2)

When adding fractions that do not have the same denominator, you must find the **lowest common denominator**. The lowest common denominator is the least common multiple of the denominators.

Example 11:   Add: $3\frac{1}{2} + 2\frac{2}{3}$

Step 1: Rewrite the problem vertically, and find the **lowest common denominator**. The lowest common denominator must be the same.
Think: What is the smallest number I can divide 2 and 3 into without a remainder? 6, of course. Recall, 6 is the lowest common multiple of 2 and 3.

$$\begin{array}{rl} 3\frac{1}{2} &= \frac{}{6} \\ +2\frac{2}{3} &= \frac{}{6} \end{array}$$

Step 2: To find the numerator for the top fraction, think: What do I multiply 2 by to get 6? You must multiply the top and bottom numbers of the fraction by 3 to keep the fraction equal. For the bottom fraction, multiply the top and bottom number by 2.

Step 3: Add whole numbers and fractions, and simplify.

$$\begin{array}{rl} 3\frac{1}{2} &= 3\frac{3}{6} \\ +2\frac{2}{3} &= 2\frac{4}{6} \\ \hline &= 5\frac{7}{6} = 6\frac{1}{6} \end{array}$$

Add and simplify the answers. (DOK 2)

1. $3\frac{5}{9}$
    $+5\frac{2}{3}$

2. $1\frac{1}{4}$
    $+4\frac{2}{5}$

3. $3\frac{3}{4}$
    $+2\frac{3}{5}$

4. $2\frac{1}{4}$
    $+1\frac{7}{8}$

5. $6\frac{5}{6}$
    $+4\frac{1}{3}$

6. $9\frac{1}{5}$
    $+5\frac{5}{6}$

7. $\frac{1}{3}$
    $+7\frac{3}{4}$

8. $9\frac{4}{9}$
    $+3\frac{2}{3}$

9. $4\frac{7}{10}$
    $+8\frac{2}{3}$

10. $5\frac{2}{7}$
    $+\frac{1}{2}$

11. $3\frac{3}{11}$
    $+2\frac{3}{4}$

12. $4\frac{3}{5}$
    $+\frac{4}{9}$

Chapter 5 Adding and Subtracting Fractions

5.9 Subtracting Mixed Numbers (DOK 2)

Example 12: Subtract: $7\frac{1}{4} - 5\frac{5}{6}$

Step 1: Rewrite the problem and find a common denominator.

$$7\frac{1}{4} \quad \genfrac{}{}{0pt}{}{\times 3}{\times 3} \rightarrow 7\frac{3}{12}$$

$$-5\frac{5}{6} \quad \genfrac{}{}{0pt}{}{\times 2}{\times 2} \rightarrow -5\frac{10}{12}$$

Refer to the "About AR" on page xi!

Step 2: You cannot subtract $\frac{10}{12}$ from $\frac{3}{12}$. You must borrow 1 from the 7. The 1 will be in the fraction form $\frac{12}{12}$ which you must add to the $\frac{3}{12}$ you already have, making $\frac{15}{12}$. Subtract whole numbers and simplify.

You must borrow → $7\overset{615}{\cancel{}\frac{\cancel{3}}{12}}$ Add 3 and 12 to get 15.

$\phantom{\text{You must borrow → }}-5\frac{10}{12}$

$\phantom{\text{You must borrow → }}\overline{1\frac{5}{12}}$

Subtract and simplify. (DOK 2)

1. $4\frac{1}{3}$
   $-1\frac{5}{9}$

2. $3\frac{4}{9}$
   $-2\frac{5}{6}$

3. $8\frac{4}{7}$
   $-5\frac{1}{3}$

4. $5\frac{2}{5}$
   $-3\frac{1}{2}$

5. $8\frac{2}{5}$
   $-5\frac{3}{10}$

6. $9\frac{2}{5}$
   $-4\frac{3}{4}$

7. $9\frac{3}{4}$
   $-2\frac{1}{3}$

8. $5\frac{1}{7}$
   $-\frac{2}{3}$

9. $6\frac{1}{5}$
   $-3\frac{3}{8}$

10. $6\frac{5}{6}$
    $-3\frac{4}{5}$

11. $2\frac{2}{9}$
    $-1\frac{3}{4}$

12. $4\frac{7}{10}$
    $-3\frac{1}{3}$

13. $7\frac{3}{5}$
    $-4\frac{5}{6}$

14. $9\frac{3}{8}$
    $-5\frac{1}{2}$

15. $8\frac{1}{9}$
    $-5\frac{1}{3}$

16. $5\frac{1}{6}$
    $-1\frac{2}{3}$

17. $6\frac{5}{6}$
    $-3\frac{1}{3}$

18. $7\frac{2}{3}$
    $-3\frac{5}{6}$

19. $8\frac{4}{7}$
    $-4\frac{3}{4}$

20. $9\frac{3}{4}$
    $-1\frac{1}{5}$

5.10 Changing Mixed Numbers to Improper Fractions (DOK 1, 2)

Example 13: Change $4\frac{3}{5}$ to an improper fraction.

Step 1: Multiply the whole number (4) by the bottom number of the fraction (5). $4 \times 5 = 20$

Step 2: Add the top number to the product from Step 1: $20 + 3 = 23$

Step 3: Put the answer from step 2 over the bottom number (5).

2. Add this number.   3. Put the answer here.

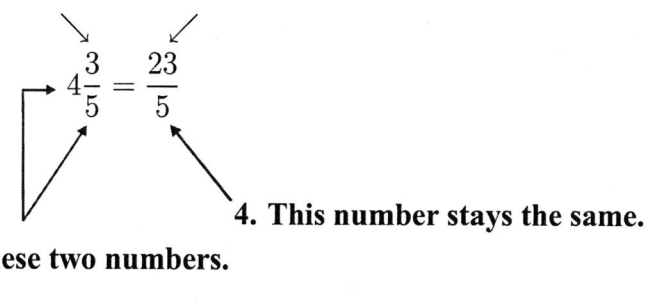

$$4\frac{3}{5} = \frac{23}{5}$$

4. This number stays the same.

1. Multiply these two numbers.

Change the following mixed numbers to improper fractions. (DOK 2)

1. $3\frac{1}{2} =$ _____
2. $2\frac{7}{8} =$ _____
3. $9\frac{2}{3} =$ _____
4. $4\frac{3}{5} =$ _____
5. $7\frac{1}{4} =$ _____

6. $8\frac{5}{8} =$ _____
7. $1\frac{2}{3} =$ _____
8. $2\frac{1}{2} =$ _____
9. $6\frac{1}{5} =$ _____
10. $5\frac{2}{5} =$ _____

11. $3\frac{3}{5} =$ _____
12. $9\frac{3}{8} =$ _____
13. $10\frac{4}{5} =$ _____
14. $3\frac{3}{10} =$ _____
15. $4\frac{1}{8} =$ _____

16. $2\frac{5}{6} =$ _____
17. $7\frac{3}{5} =$ _____
18. $6\frac{7}{8} =$ _____
19. $7\frac{2}{5} =$ _____
20. $7\frac{2}{3} =$ _____

Whole numbers become improper fractions when you put them over 1. Change the following whole numbers to improper fractions. The first one is done for you. (DOK 1)

21. $4 = \frac{4}{1}$
22. $10 =$ _____
23. $3 =$ _____
24. $2 =$ _____
25. $15 =$ _____
26. $5 =$ _____
27. $6 =$ _____
28. $11 =$ _____
29. $8 =$ _____
30. $16 =$ _____

Chapter 5   Adding and Subtracting Fractions

5.11   Estimating Fractions (DOK 2)

Estimating fractions can be done using mental math.

Example 14:   *Estimate* the sum of $\frac{1}{3}$ and $\frac{1}{4}$ to the nearest half.

Step 1:   Look carefully at both fractions. Notice that both are less than $\frac{1}{2}$, therefore you know that added together, the sum will be less than one.

Step 2:   Now narrow it down a bit more. The fraction $\frac{1}{3}$ is a little more than $\frac{1}{4}$, so if you added the two fractions together in your head, you know that they add up to a little over $\frac{1}{2}$ (as long as you know that $\frac{1}{4} + \frac{1}{4} = \frac{1}{2}$).

Answer: $\frac{1}{3} + \frac{1}{4} =$ a little more than $\frac{1}{2}$.

Using addition and subtraction for the fraction problems below, estimate the answers by putting an √ under one of the 4 choices given. The first one is done for you. (DOK 2)

	Between 0 and $\frac{1}{2}$	Between $\frac{1}{2}$ and 1	Between 1 and $1\frac{1}{2}$	Between $1\frac{1}{2}$ and 2
1. $\frac{5}{8} + \frac{4}{8}$			√	
2. $\frac{7}{8} - \frac{1}{8}$				
3. $1\frac{1}{8} + \frac{1}{8}$				
4. $2\frac{11}{12} - 2\frac{1}{12}$				
5. $\frac{5}{6} + \frac{2}{6}$				
6. $\frac{3}{4} - \frac{2}{4}$				
7. $\frac{8}{12} + \frac{1}{12}$				
8. $\frac{5}{6} - \frac{1}{6}$				
9. $1\frac{5}{8} + \frac{1}{8}$				
10. $7\frac{4}{5} - 6\frac{1}{5}$				

5.11 Estimating Fractions (DOK 2)

Example 15: *Estimate* the sum of $4\frac{1}{8} + 6\frac{5}{6} + 2\frac{1}{4}$, to the nearest whole number.

Step 1: Round each fraction to the nearest whole number - zero or one. A fraction that is less than $\frac{1}{2}$ will round down to zero, and a fraction that is $\frac{1}{2}$ or greater will round up to one.
$\frac{1}{8}$ rounds down to zero, $\frac{5}{6}$ rounds up to one, and $\frac{1}{4}$ rounds down to zero.

Step 2: Add up the rounded fractions: zero + zero + one = one

Step 3: Add the total of the rounded fractions to the whole numbers:
$4 + 6 + 2 + 1 = 13$

Answer: $4\frac{1}{8} + 6\frac{5}{6} + 2\frac{1}{4}$ is *about* 13.

Estimate the sum of the mixed numbers below to the nearest whole number. (DOK 2)

1. $6\frac{1}{5} + 1\frac{2}{3} + 3\frac{7}{8}$ _____

2. $8\frac{41}{100} + 3\frac{1}{4} + 2\frac{3}{8}$ _____

3. $1\frac{1}{8} + 4\frac{73}{100} + 2\frac{1}{5}$ _____

4. $10\frac{1}{3} + 7\frac{1}{8} + 9\frac{5}{6}$ _____

5. $7\frac{1}{2} + 1\frac{1}{6} + 3\frac{3}{4}$ _____

6. $6\frac{3}{4} + 1\frac{1}{8} + 4\frac{53}{100}$ _____

7. $11\frac{1}{4} + 20\frac{3}{4} + 2\frac{3}{8}$ _____

8. $5\frac{1}{2} + 1\frac{3}{4} + 6\frac{1}{8}$ _____

9. $50\frac{2}{3} + 1\frac{251}{1,000} + 1\frac{5}{6}$ _____

10. $1\frac{5}{8} + 1\frac{1}{4} + 1\frac{1}{8}$ _____

11. $7\frac{1}{3} + 1\frac{3}{4} + 2\frac{2}{3}$ _____

12. $4\frac{1}{5} + 2\frac{1}{4} + 1\frac{13}{100}$ _____

13. $1\frac{1}{100} + 1\frac{1}{1,000} + 1\frac{99}{100}$ _____

14. $2\frac{5}{6} + 2\frac{1}{4} + 2\frac{7}{8}$ _____

15. $8\frac{7}{100} + 8\frac{1}{4} + 8\frac{391}{1,000}$ _____

Choose the best answer for the estimated problems below. (DOK 2)

16. There were 87 customers at the food court at noon, on Saturday. About $\frac{2}{3}$ of the customers chose chicken fingers for their lunch. Which number is <u>closest</u> to the number of chicken finger eating customers?

    A) 50   B) 60   C) 70   D) 80

17. Out of 10,000 books at the town library, about $\frac{1}{10}$ of the books are classified as fiction. Which number is <u>closest</u> to the actual number of books classified as fiction?

    A) 982   B) 1,502   C) 1,999   D) 2,448

18. There are 842 students at Blue Sky Elementary. Of these, $\frac{1}{4}$ of the students chose salad for lunch on Tuesday. Which number is <u>closest</u> to the actual number of students who chose salad on Tuesday?

    A) 520   B) 410   C) 320   D) 210

Chapter 5 Adding and Subtracting Fractions

5.12 Comparing Fractions (DOK 2)

Comparing two fractions means finding which fraction is larger or smaller than the other.

Example 16: Compare $\frac{3}{4}$ and $\frac{5}{8}$. Use > or < to show their relationship.

Step 1: Write the two fractions. $\quad \frac{3}{4} \quad \frac{5}{8}$

Step 2: Multiply from the upper left to lower right and write down your answer.
$3 \times 8 = 24$
$$\frac{3}{4} \searrow \frac{5}{8}$$

Step 3: Multiply from the lower left to upper right and write down your answer after the first answer.
$4 \times 5 = 20$
$$\frac{3}{4} \nearrow \frac{5}{8}$$

Step 4: Determine which sign, > or <, would fit between 24 and 20.
$24 > 20$
$$\frac{3}{4} \bowtie \frac{5}{8}$$

This is the sign to use between the fractions.

$\frac{3}{4} > \frac{5}{8} \quad \frac{3}{4}$ is greater than $\frac{5}{8}$.

Fill in the box with the correct sign (>, <, or =). (DOK 2)

1. $\frac{3}{5} \square \frac{3}{4}$

2. $\frac{6}{8} \square \frac{5}{6}$

3. $\frac{4}{6} \square \frac{4}{5}$

4. $\frac{3}{10} \square \frac{4}{12}$

5. $\frac{5}{8} \square \frac{4}{10}$

6. $\frac{5}{8} \square \frac{4}{5}$

7. $\frac{9}{10} \square \frac{8}{12}$

8. $\frac{2}{100} \square \frac{1}{10}$

9. $\frac{4}{5} \square \frac{3}{5}$

10. $\frac{2}{6} \square \frac{4}{5}$

11. $\frac{7}{12} \square \frac{6}{10}$

12. $\frac{3}{12} \square \frac{5}{100}$

5.13 Fraction Word Problems (DOK 2)

Example 17: Beth's kittens weigh $1\frac{1}{10}$ lb, $1\frac{3}{5}$ lb, $1\frac{2}{5}$ lb, and $1\frac{3}{10}$ lb. What is the sum of the weights of her four kittens?

Step 1: Change the fractions in the problem so they all have a common denominator. The denominators are 5 and 10. It will be easiest to use 10. $1\frac{1}{10}, 1\frac{3}{5} = 1\frac{6}{10}, 1\frac{2}{5} = 1\frac{4}{10}$, and $1\frac{3}{10}$.

Step 2: Add the fractions. $1\frac{1}{10} + 1\frac{6}{10} + 1\frac{4}{10} + 1\frac{3}{10} = 4\frac{14}{10}$.

Step 3: Simplify the answer to lowest terms. $4\frac{14}{10} = 5\frac{4}{10} = 5\frac{2}{5}$

Answer: $5\frac{2}{5}$ lb

Solve and simplify answers to lowest terms. (DOK 2)

1. Jeff chews on a piece of bubble gum for $2\frac{1}{4}$ hours. Then he chews a piece of fruit gum for $1\frac{2}{3}$ hours. How long did Jeff chew the two pieces of gum? _____

2. Juan spent $3\frac{4}{8}$ hours reading from two books. One book he read for $1\frac{1}{4}$ hours. How long did he read the other book? _____

3. Marie gives her puppy a bath and uses $5\frac{4}{6}$ gallons of water. She and the puppy splash away $3\frac{2}{3}$ gallons of the dirty water. How much water does she have left? _____

4. Beau caught 3 fish weighing $4\frac{1}{4}$ lb, $5\frac{2}{8}$ lb, and $3\frac{3}{8}$ lb. How much did the three fish weigh, altogether? _____

5. Mrs. Arkin was making cookies to sell at the school bake sale. She started with $18\frac{1}{2}$ cups of oats in the oat canister. The first batch of cookies used $2\frac{1}{4}$ cups, the second batch used $3\frac{1}{8}$ cups, and the third batch used 3 cups. How many cups of oats are left in the oat canister? _____

6. Alicia has a piece of ribbon measuring $5\frac{3}{4}$ feet. She cuts off a $2\frac{1}{8}$ foot piece of the ribbon. What length of ribbon remains? _____

7. Zachary worked $6\frac{1}{4}$ hours at the school car wash fund raiser. Jeremy worked $7\frac{7}{8}$ hours at the same fund raiser. How many more hours did Jeremy work than Zachary? _____

8. Mr. Garcia brought home 3 large pizzas for a pizza birthday party. Mr. Garcia ate $\frac{1}{2}$ of one pizza. Mrs. Garcia ate $\frac{1}{4}$ of one pizza. The kids and guests ate $2\frac{1}{8}$ pizzas. The dog took $\frac{1}{8}$ of a pizza. Is there any pizza left over? If so, how much? _____

9. The Williams family is planting a garden. Mr. Williams planted $\frac{1}{3}$ of the garden, Mrs. Williams planted $\frac{1}{6}$ of the garden. The children planted the remainder of the garden. What fraction of the garden did the children plant? _____

Chapter 5   Adding and Subtracting Fractions

5.14   Modeling Fractions (DOK 2)

Use the fraction strips below and add or subtract as required. Simplify your answer if needed. (DOK 2)

1. $\frac{1}{4} + \frac{1}{8}\frac{1}{8}\frac{1}{8}$

2. $\frac{1}{8}\frac{1}{8}\frac{1}{8}\frac{1}{8}\frac{1}{8} + \frac{1}{3}$

3. $\frac{1}{2} + \frac{1}{2} + \frac{1}{3}\frac{1}{3} - \frac{1}{4}\frac{1}{4}\frac{1}{4}$

Use the shaded parts of the area models below and add or subtract as required. Simplify your answer if needed.

4.

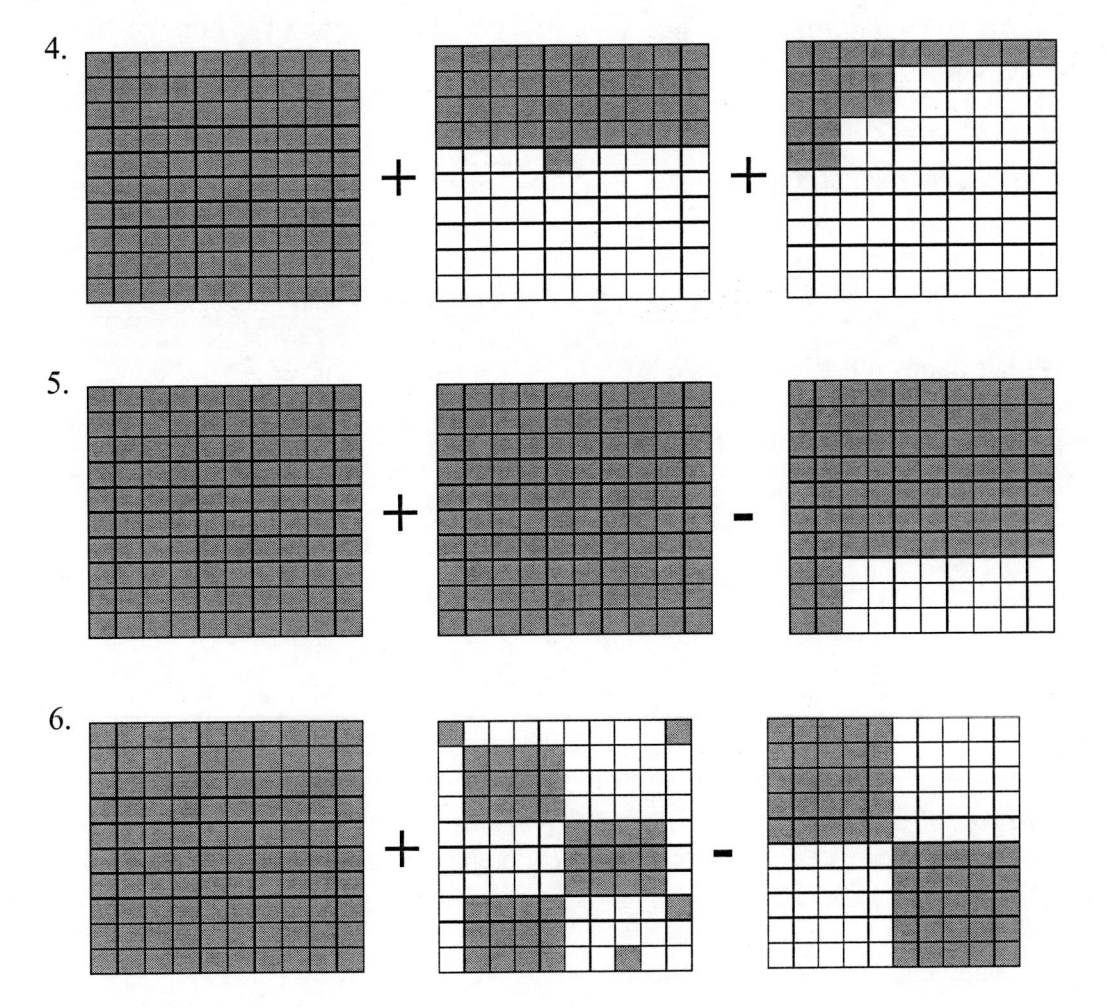

5.

6.

5.15 Going Deeper into Adding and Subtracting Fractions (DOK 3)

Solve the multi-step problems below. Show your work for each step.

1. There are $2\frac{1}{2}$ pizzas left over from a pizza party. Aaron takes home $\frac{3}{4}$ of one pizza, Bert takes home $\frac{5}{8}$ of one pizza, and Lenny takes home the rest of the pizza. How much pizza does Lenny <u>and</u> Bert take home?

2. Mrs. Anderson is a fifth grade teacher that has her students tracking how many books they read for the school year. The table below shows the number of books read by the students in row one of her classroom.

Student	Number of Books Read
John	$3\frac{1}{4}$
Aria	$4\frac{3}{8}$
Jim	$2\frac{7}{8}$
Maria	$5\frac{1}{2}$
Jacob	$4\frac{3}{4}$
Reagan	$3\frac{1}{2}$

    Part 1: How many books have the students read in all?

    Part 2: How many more books did Aria read than Reagan?

    Part 3: How many less books did Jim read than Maria?

3. In the pantry cupboard of the Wilcox family there are 3 kinds of cereal. One kind of cereal has $\frac{2}{5}$ of a box left, a second kind has $\frac{3}{4}$ of a box left, and a third kind has $1\frac{4}{5}$ boxes left.

    Part 1: How many boxes of cereal are left in all?

    Part 2: If Evan and his little sister eat $\frac{3}{5}$ of a box in all for breakfast, how much cereal remains in the pantry?

4. Which of the following pairs of numbers has a LCM of 24?

    3 and 9   4 and 5   3 and 8

5. Mark has $24\frac{3}{5}$ dollars in his pocket. He has $16\frac{2}{10}$ dollars in his piggy bank. He spends $18\frac{4}{5}$ dollars on 1 video game. His father gives him his allowance of $4\frac{1}{2}$ dollars. He gives his mother $2\frac{1}{2}$ dollars he owes her from a shopping trip. How much money does Mark have left in all?

6. A mother cat was hiding her litter behind the washing machine at home. She is busy moving her litter to a new hiding place behind the couch. So far she has moved $\frac{2}{5}$ of the kittens. She has $\frac{1}{5}$ of the kittens in her mouth. What fraction of the kittens are still behind the washing machine?

7. It's the last round of a math quiz bowl game. Your team is the Matherines and the problem you are asked to solve is shown below.

    $$4\frac{1}{4} + 3\frac{1}{3} + 2\frac{1}{2} - 1\frac{1}{12} - 2\frac{1}{4} - 3\frac{1}{6}$$

    Find the answer of this problem that includes addition and subtraction.

8. The Math-tas-tics are in the same quiz bowl. They are given a similar problem:

    $$5\frac{3}{5} + 1\frac{1}{5} + 6\frac{3}{10} - 7\frac{1}{10} - 2\frac{3}{10} - 1\frac{4}{5} = 2\frac{1}{5}$$

    The Math-tas-tics' team got the answer wrong. Give the correct answer.

Chapter 5 Adding and Subtracting Fractions

Chapter 5 Review

Simplify. (DOK 1)

1. $\dfrac{2}{6}$ _____
2. $\dfrac{15}{6}$ _____
3. $\dfrac{22}{4}$ _____
4. $\dfrac{123}{30}$ _____
5. $\dfrac{3}{24}$ _____
6. $\dfrac{6}{36}$ _____
7. $\dfrac{7}{56}$ _____
8. $\dfrac{15}{45}$ _____

Change to an improper fraction. (DOK 2)

9. $114\frac{1}{2}$ _____
10. $8\frac{2}{3}$ _____
11. $10\frac{4}{5}$ _____
12. $11\frac{1}{4}$ _____

Add and simplify. (DOK 2)

13. $\frac{2}{5} + \frac{3}{5}$ _____
14. $2\frac{1}{4} + 7\frac{1}{4}$ _____
15. $3\frac{1}{15} + \frac{3}{15}$ _____
16. $\frac{1}{8} + \frac{3}{8}$ _____
17. $10\frac{5}{6} + 9\frac{1}{6}$ _____
18. $13\frac{7}{8} + 8\frac{1}{2}$ _____

Subtract and simplify. (DOK 2)

19. $12\frac{3}{4} - 6\frac{1}{4}$ _____
20. $8\frac{4}{5} - \frac{3}{5}$ _____
21. $841\frac{3}{8} - 14\frac{1}{8}$ _____
22. $10\frac{4}{5} - 1\frac{3}{5}$ _____
23. $11\frac{1}{5} - 5\frac{4}{5}$ _____
24. $24\frac{1}{5} - 14\frac{5}{6}$ _____

Use the >, <, and = signs to make the following correct. (DOK 2)

25. $\dfrac{2}{3}\ \square\ \dfrac{2}{6}$
26. $\dfrac{4}{8}\ \square\ \dfrac{1}{2}$
27. $\dfrac{1}{100}\ \square\ \dfrac{1}{1,000}$
28. $\dfrac{1}{10}\ \square\ \dfrac{1}{6}$

Chapter 5 Review

Read each word problem and solve. (DOK 2)

29. Isabelle knows that each quarter she has is equal to $\frac{1}{4}$ of a dollar. She has 17 quarters. How many dollars does she have? (Include any fraction of a dollar she may have.) _____

30. Joshua is $61\frac{1}{2}$ inches tall. He hopes to be as tall as his father when he grows up. His father is $74\frac{1}{4}$ inches tall. How many more inches does Joshua need to grow to be as tall as his father? _____

31. Harold knew he had about three and a half dollars. Which of the following is closest to three and a half dollars?
    $3\frac{1}{4}$ $3\frac{3}{8}$ $3\frac{3}{4}$ $3\frac{7}{8}$

Solve the multi-step problems below. Show your work for each step. (DOK 3)

32. $16\frac{4}{5} + 12\frac{3}{10} + 1\frac{1}{5} - 11\frac{7}{10} - 5\frac{1}{5}$

33. $2\frac{3}{4} + 9\frac{5}{12} + 7\frac{5}{6} - 3\frac{7}{12} - 4\frac{1}{6}$

Find the missing numerators from the following equivalent fractions. (DOK 2)

34. $\frac{1}{6} = \frac{\ }{12}$

35. $\frac{2}{5} = \frac{\ }{25}$

36. $\frac{1}{8} = \frac{\ }{24}$

37. $\frac{3}{5} = \frac{\ }{15}$

38. $\frac{8}{10} = \frac{\ }{100}$

39. $\frac{2}{6} = \frac{\ }{36}$

Use the fraction strips and area models below to add or subtract as required. Simplify your answer if needed. (DOK 2)

40. $\boxed{\frac{1}{4}}\ \boxed{\frac{1}{4}} + \boxed{\frac{1}{3}}$

41.

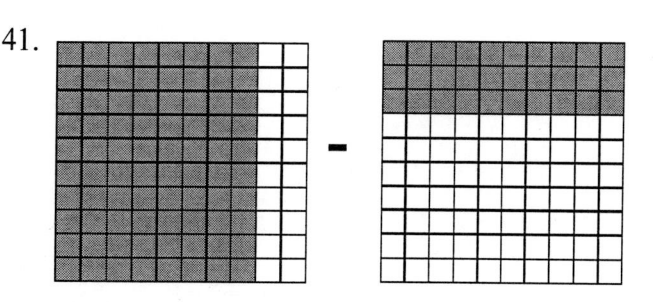

Chapter 5 Adding and Subtracting Fractions

Chapter 5 Test

1 Simplify the improper fraction $\frac{31}{2}$.

A $15\frac{1}{5}$

B $15\frac{1}{2}$

C $15\frac{3}{4}$

D $15\frac{7}{8}$

(DOK 1)

2 Convert $2\frac{2}{3}$ to an improper fraction.

A $\frac{8}{3}$

B $\frac{7}{3}$

C $\frac{6}{3}$

D $\frac{4}{3}$

(DOK 2)

3 Add: $\frac{1}{6} + \frac{4}{6} =$

A $\frac{1}{6}$

B $\frac{3}{6}$

C $\frac{5}{6}$

D $1\frac{2}{6}$

(DOK 1)

4 Subtract: $6\frac{2}{3} - 3\frac{1}{3} =$

A $2\frac{1}{3}$

B 3

C $3\frac{1}{3}$

D $3\frac{1}{2}$

(DOK 1)

5 Subtract: $346\frac{1}{4} - 17\frac{3}{4} =$

A $328\frac{1}{4}$

B $328\frac{1}{2}$

C $329\frac{1}{4}$

D $329\frac{1}{2}$

(DOK 2)

6 Which fraction comparison is not true?

A $\frac{1}{2} > \frac{1}{4}$

B $\frac{2}{3} < \frac{3}{4}$

C $\frac{1}{2} > \frac{2}{3}$

D $\frac{3}{4} > \frac{1}{2}$

(DOK 2)

7 John Paul has read $\frac{2}{5}$ of a novel. Which fraction is equal to $\frac{2}{5}$?

A $\frac{2}{10}$

B $\frac{4}{10}$

C $\frac{3}{5}$

D $\frac{4}{5}$

(DOK 1)

Chapter 5 Test

8 Which improper fraction is equal to $14\frac{3}{4}$?

A $\frac{16}{4}$

B $\frac{21}{4}$

C $\frac{17}{4}$

D $\frac{59}{4}$

(DOK 2)

9 Dr. Sheldon, a veterinarian, weighed Carl's 3 cats. Bootsie weighed $7\frac{1}{8}$ pounds, Bitsy weighed $6\frac{5}{8}$ pounds, and Boo weighed $7\frac{1}{8}$ pounds. What is the total weight of the 3 cats?

A $22\frac{5}{8}$ lb

B $21\frac{7}{8}$ lb

C $20\frac{7}{8}$ lb

D $19\frac{7}{8}$ lb

(DOK 2)

10 Solve: $3\frac{1}{2} + 14\frac{1}{12} - 8\frac{3}{4} - 1\frac{1}{6}$

A $10\frac{1}{3}$

B $10\frac{2}{3}$

C $10\frac{5}{12}$

D $7\frac{2}{3}$

(DOK 3)

11 Which mixed number is equal to $\frac{37}{5}$?

A $7\frac{2}{5}$

B $6\frac{3}{5}$

C $7\frac{3}{5}$

D $6\frac{2}{5}$

(DOK 2)

12 Solve: $7\frac{1}{5} + 10\frac{7}{10} - 2\frac{9}{10} - 12\frac{4}{5}$

A $2\frac{1}{5}$

B $2\frac{1}{10}$

C $3\frac{1}{10}$

D $3\frac{3}{5}$

(DOK 3)

13 Add the fraction strips below.

$\boxed{\dfrac{1}{2}} + \boxed{\dfrac{1}{8}}\boxed{\dfrac{1}{8}}\boxed{\dfrac{1}{8}}$

A $\frac{3}{4}$

B $\frac{5}{8}$

C $\frac{7}{8}$

D $\frac{4}{16}$

(DOK 2)

14 Ethan has a length of rope measuring $37\frac{7}{8}$ feet. He uses $28\frac{1}{2}$ feet to make a swing to put on the big oak tree in his back yard. How much rope does Ethan have leftover?

A $10\frac{7}{8}$

B $10\frac{1}{4}$

C $9\frac{3}{8}$

D $9\frac{1}{4}$

(DOK 2)

Copyright © American Book Company

Chapter 5   Adding and Subtracting Fractions

15 Which fraction is equal to $\frac{3}{4}$?

  A $\frac{6}{4}$

  B $\frac{6}{8}$

  C $\frac{2}{3}$

  D $\frac{4}{3}$

(DOK 2)

16 What is the least common multiple of 6 and 8?

  A 6
  B 8
  C 12
  D 24

(DOK 2)

17 The sum of $\frac{2}{3} + \frac{1}{4}$ is

  A in between $\frac{1}{2}$ and 1.
  B in between 1 and $1\frac{1}{2}$.
  C in between $1\frac{1}{2}$ and 2.
  D in between 2 and $2\frac{1}{2}$.

(DOK 2)

18 Add: $3\frac{1}{4} + 6\frac{1}{5}$.

  A $9\frac{9}{20}$

  B $9\frac{2}{9}$

  C $9\frac{1}{9}$

  D $9\frac{9}{10}$

(DOK 2)

19 Subtract: $11\frac{5}{6} - 4\frac{2}{3}$.

  A $7\frac{3}{3}$

  B $7\frac{1}{6}$

  C $7\frac{7}{9}$

  D $7\frac{1}{3}$

(DOK 2)

20 What is the least common multiple of 7 and 10?

  A 35
  B 14
  C 70
  D 63

(DOK 2)

21 Simplify: $\frac{18}{5}$.

  A $4\frac{2}{5}$

  B $4\frac{3}{5}$

  C $3\frac{2}{5}$

  D $3\frac{3}{5}$

(DOK 2)

22 Which fraction comparison is true?

  A $\frac{2}{3} < \frac{1}{4}$

  B $\frac{7}{8} < \frac{3}{4}$

  C $\frac{3}{4} > \frac{2}{3}$

  D $\frac{1}{4} > \frac{1}{3}$

(DOK 2)

Chapter 6
Multiplying Fractions

This chapter covers the following CC 5 standards:

Number and Operations - Fractions	5.NF.4, 5.NF.5, 5.NF.6

6.1 Multiplying Whole Numbers by Fractions (DOK 2)

When a whole number is multiplied by a fraction, the product will be **less** than the whole number. It is like asking for a **part**, or a fraction, of a whole number. For instance, you may be given two-thirds of a package of gum. The package contains 15 sticks of gum. Two-thirds of the package is 10 sticks of gum: $15 \times \frac{2}{3} = \frac{30}{3}$, simplified, is 10. Notice that the whole number is multiplied by the numerator, then that product is divided by the denominator.

Reminder: A fraction is made of the numerator over the denominator: $\frac{\text{numerator}}{\text{denominator}}$

Example 1: Multiply $18 \times \frac{5}{6}$.

    Step 1: 18 is the same as $\frac{18}{1}$. Put 18 over 1. $\frac{18}{1} \times \frac{5}{6}$. Cancel where you can.

        $\frac{3\cancel{18}}{1} \times \frac{5}{\cancel{6}1}$

    Step 2: Multiply across. $\frac{3}{1} \times \frac{5}{1} = \frac{3 \times 5}{1 \times 1} = \frac{15}{1} = 15$

    Answer: 15

Example 2: Abigail has completed $\frac{1}{3}$ of a book that has 243 pages in it. How many pages of the book has Abigail read?

    Step 1: Remember "of" means multiply. 243 is the same as $\frac{243}{1}$. Put 243 over 1. $\frac{243}{1} \times \frac{1}{3}$. Cancel where you can. $\frac{81\cancel{243}}{1} \times \frac{1}{\cancel{3}1}$

    Step 2: Multiply across. $\frac{81}{1} \times \frac{1}{1} = \frac{81 \times 1}{1 \times 1} = \frac{81}{1} = 81$

    Answer: Abigail has read 81 pages of the book.

Copyright © American Book Company

Chapter 6   Multiplying Fractions

Multiply the whole numbers by the fractions. (DOK 2)

1. $24 \times \frac{1}{2}$ _____
2. $36 \times \frac{5}{6}$ _____
3. $15 \times \frac{1}{3}$ _____
4. $60 \times \frac{5}{12}$ _____
5. $81 \times \frac{2}{9}$ _____

6. $99 \times \frac{1}{3}$ _____
7. $10 \times \frac{2}{5}$ _____
8. $160 \times \frac{3}{4}$ _____
9. $20 \times \frac{3}{5}$ _____
10. $6 \times \frac{1}{3}$ _____

11. $22 \times \frac{3}{11}$ _____
12. $34 \times \frac{1}{2}$ _____
13. $45 \times \frac{2}{5}$ _____
14. $77 \times \frac{3}{11}$ _____
15. $65 \times \frac{4}{5}$ _____

16. $48 \times \frac{3}{8}$ _____
17. $30 \times \frac{5}{6}$ _____
18. $64 \times \frac{1}{8}$ _____
19. $27 \times \frac{5}{9}$ _____
20. $42 \times \frac{2}{7}$ _____

21. Jeremy used $\frac{3}{4}$ of a tablet of drawing paper so far this month. The tablet started with 80 pages. How many pages of drawing paper has Jeremy used so far this month?

22. Mr. Barton has $\frac{5}{8}$ of a tank of gas left in his car. The car's gas tank holds 16 gallons. How many gallons of gas does Mr. Barton have left in his gas tank?

Example 3:   Model: $\frac{1}{4} \times 3 = \frac{1}{4} \times \frac{3}{1} = \frac{3}{4}$.

   Step 1:   Divide a circle into 4 equal parts.

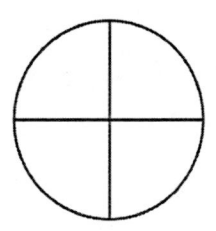

   Step 2:   Shade 3 of the 4 parts. This model shows $\frac{3}{4}$.

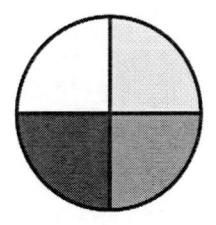

6.1   Multiplying Whole Numbers by Fractions (DOK 2)

Match the fraction problems on the left column to the models on the right column. (DOK 2)

1. $4 \times \dfrac{1}{6}$

2. $3 \times \dfrac{1}{2}$

3. $5 \times \dfrac{2}{3}$

4. $6 \times \dfrac{1}{3}$

5. $2 \times \dfrac{5}{8}$

6. $7 \times \dfrac{1}{4}$

7. $5 \times \dfrac{3}{8}$

8. $4 \times \dfrac{1}{3}$

A.

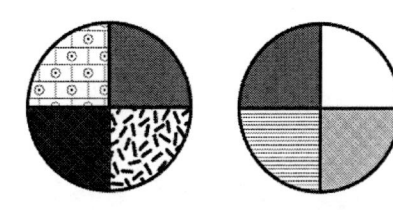

B.

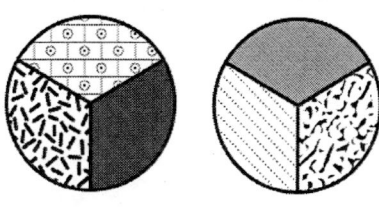

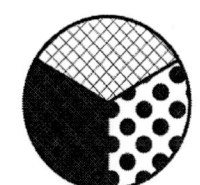

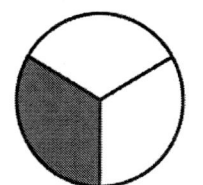

C.

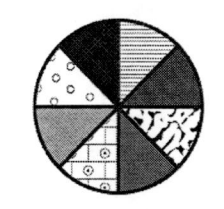

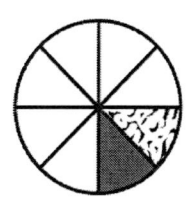

D.

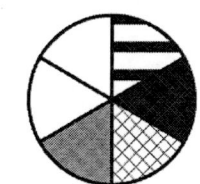

E.

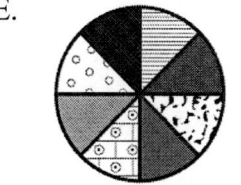

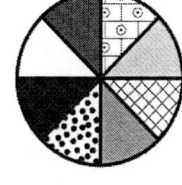

F.

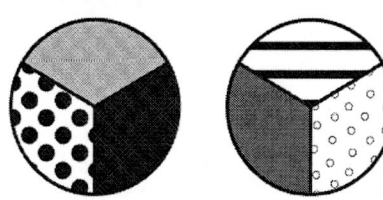

G.

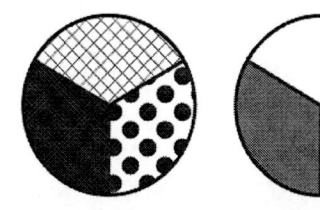

H.

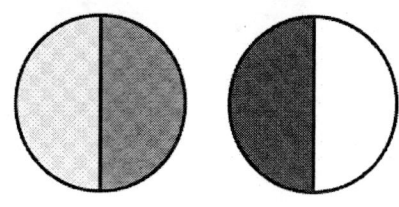

Copyright © American Book Company

79

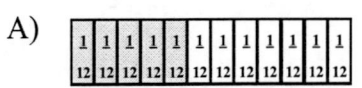

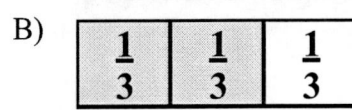

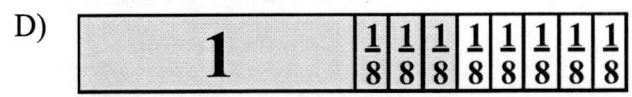

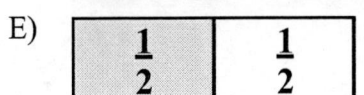

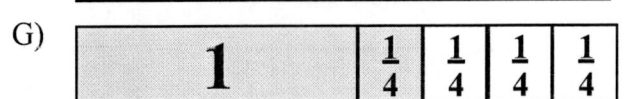

Chapter 6 Multiplying Fractions

6.2 More Multiplying Whole Numbers by Fractions (DOK 2)

The fraction model below shows $\frac{7}{8}$ of 1, shaded. One is re-sized into 8 parts, 7 of which are shaded.

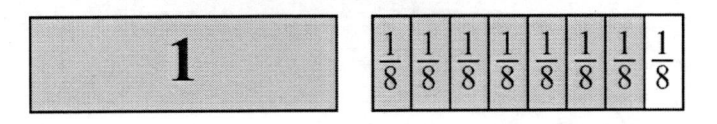

This shaded fraction model shows $1\frac{1}{2}$: the sum of $1 + (3 \times \frac{1}{6}) = 1\frac{3}{6} = 1\frac{1}{2}$.

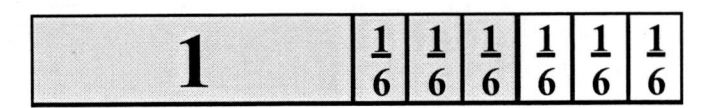

Solve the fraction multiplication problems on the left column, simplify, and match them to the shaded fraction model on the right column. Some of the fraction models may be used more than once. (DOK 2)

1. $8 \times \dfrac{1}{16} =$

2. $3 \times \dfrac{2}{9} =$

3. $7 \times \dfrac{2}{24} =$

4. $4 \times \dfrac{2}{6} =$

5. $15 \times \dfrac{1}{36} =$

6. $3 \times \dfrac{3}{12} =$

7. $5 \times \dfrac{4}{24} =$

8. $11 \times \dfrac{2}{16} =$

9. $2 \times \dfrac{5}{8} =$

10. $7 \times \dfrac{2}{28} =$

11. $5 \times \dfrac{5}{20} =$

6.3 Multiplying Fractions with Canceling (DOK 2)

Example 4: Multiply $\frac{2}{3} \times \frac{3}{4}$

Step 1: Multiply the top numbers $2 \times 3 = 6$. The top of the fraction is 6.

Step 2: Multiply the bottom numbers $3 \times 4 = 12$. The bottom of the fraction is 12.

$$\frac{2}{3} \times \frac{3}{4} = \frac{6}{12}$$

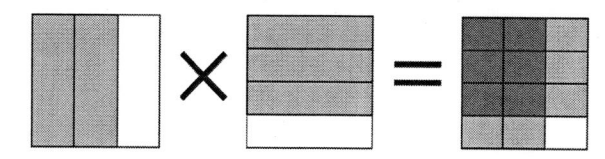

Multiply the fractions. Simplify if needed. (DOK 2)

1. $\frac{4}{7} \times \frac{3}{5}$ _____
2. $\frac{3}{4} \times \frac{1}{5}$ _____
3. $\frac{2}{3} \times \frac{1}{7}$ _____
4. $\frac{2}{3} \times \frac{1}{3}$ _____
5. $\frac{1}{5} \times \frac{4}{9}$ _____
6. $\frac{3}{10} \times \frac{1}{4}$ _____
7. $\frac{1}{2} \times \frac{1}{3}$ _____
8. $\frac{3}{4} \times \frac{1}{3}$ _____
9. $\frac{1}{9} \times \frac{1}{4}$ _____

Example 5: Multiply $\frac{2}{3} \times \frac{3}{4}$ using canceling.

Step 1: In this problem, the 3's are both divisible by 3, so they cancel. $3 \div 3 = 1$

$$\frac{2}{\cancel{3}} \times \frac{\cancel{3}^1}{4} = \frac{2}{1} \times \frac{1}{4}$$

Step 2: The 2 and the 4 are both divisible by 2, so they cancel. $2 \div 2 = 1$ and $4 \div 2 = 2$

$$\frac{\cancel{2}^1}{1} \times \frac{1}{\cancel{4}_2} = \frac{1}{1} \times \frac{1}{2}$$

Step 3: Multiply. $\frac{1}{1} \times \frac{1}{2} = \frac{1}{2}$

Cancel where possible in the following problems, then multiply. (DOK 2)

10. $\frac{2}{3} \times \frac{3}{8}$ _____
11. $\frac{3}{4} \times \frac{4}{9}$ _____
12. $\frac{3}{8} \times \frac{2}{3}$ _____
13. $\frac{5}{6} \times \frac{3}{10}$ _____
14. $\frac{2}{7} \times \frac{1}{4}$ _____
15. $\frac{5}{9} \times \frac{1}{5}$ _____
16. $\frac{8}{9} \times \frac{3}{4}$ _____
17. $\frac{9}{10} \times \frac{5}{6}$ _____
18. $\frac{3}{7} \times \frac{7}{9}$ _____
19. $\frac{6}{7} \times \frac{1}{12}$ _____
20. $\frac{2}{5} \times \frac{5}{6}$ _____
21. $\frac{4}{9} \times \frac{3}{8}$ _____

Chapter 6   Multiplying Fractions

6.4   Multiplying Mixed Numbers (DOK 2)

Example 6: Multiply $4\frac{3}{8} \times \frac{8}{10}$

Step 1: Change the mixed numbers in the problem to improper fractions. $4\frac{3}{8} = \frac{35}{8}$

Step 2: When multiplying fractions, you can cancel and simplify terms that have a common factor. The 8 in the first fraction will cancel with the 8 in the second fraction.

$$\frac{35}{\cancel{8}_1} \times \frac{\cancel{8}^1}{10}$$

The terms 35 and 10 are both divisible by 5, so

35 simplifies to 7, and 10 simplifies to 2.

$$\frac{\cancel{35}^7}{1} \times \frac{1}{\cancel{10}_2} = \frac{7 \times 1}{1 \times 2} = \frac{7}{2}$$

Refer to the "About AR" on page xi!

Step 3: Simplify the improper fraction. $\frac{7}{2}$

Step 4: You cannot leave an improper fraction as the answer, so to change $\frac{7}{2}$ back to a mixed number.

$\frac{7}{2} = 3\frac{1}{2}$

Multiply and simplify your answers to lowest terms. (DOK 2)

1. $3\frac{1}{5} \times 1\frac{1}{2}$ _____
2. $\frac{3}{8} \times 3\frac{3}{7}$ _____
3. $4\frac{1}{3} \times 2\frac{1}{4}$ _____
4. $4\frac{2}{3} \times 3\frac{3}{4}$ _____

5. $1\frac{1}{2} \times 1\frac{2}{5}$ _____
6. $3\frac{3}{7} \times \frac{5}{6}$ _____
7. $3 \times 6\frac{1}{3}$ _____
8. $1\frac{1}{6} \times 8$ _____

9. $6\frac{2}{5} \times 5$ _____
10. $6 \times 1\frac{3}{8}$ _____
11. $\frac{5}{7} \times 2\frac{1}{3}$ _____
12. $1\frac{2}{5} \times 1\frac{1}{4}$ _____

13. $2\frac{1}{2} \times 5\frac{4}{5}$ _____
14. $7\frac{2}{3} \times \frac{3}{4}$ _____
15. $2 \times 3\frac{1}{4}$ _____
16. $3\frac{1}{8} \times 1\frac{3}{5}$ _____

6.5 Interpreting Multiplication (DOK 2)

When one factor is a whole number:

At a 5k race, cases of bottled water were brought in for the runners. After the race, there were 3 partial cases of water left. Each partial case had $\frac{2}{3}$ of the bottles of water left.

$3 \times \frac{2}{3} = \frac{6}{3} = 2$. There are 2 cases of water left over.

As seen in the problem above, when you multiply a whole number by a fraction **less** than 1, the product is **less** than the whole number.

At another 5k race, there were five water stations. After the race, each station had $\frac{5}{4}$ $\left(1\frac{1}{4}\right)$ cases left over.

$5 \times \frac{5}{4} = \frac{25}{4} = 6\frac{1}{4}$. There are $6\frac{1}{4}$ cases of water left over.

As seen in this problem, when you multiply a whole number by a fraction **greater** than 1, the product is **more** than the whole number.

When both factors are fractions:

$\frac{2}{5} \times \frac{1}{2} = \frac{2}{10} = \frac{1}{5}$

$\frac{2}{3} \times \frac{6}{5} = \frac{12}{15} = \frac{4}{5}$

The original fraction will be <u>less</u> if multiplied by a number <u>less</u> than 1.
The original fraction will be <u>greater</u> if multiplied by a number <u>greater</u> than 1.

However, if both factors are **greater** than one, the product will be **greater** than one.

$\frac{4}{3} \times \frac{7}{5} = \frac{28}{15} = 1\frac{13}{15}$

Look carefully at each problem below. Determine if the answer will be less than or more than one. Write <u>less</u> or <u>more</u> for your answers. (DOK 2)

1. $\frac{2}{5} \times 4$ _____
2. $\frac{1}{3} \times \frac{6}{7}$ _____
3. $\frac{4}{2} \times \frac{9}{5}$ _____
4. $8 \times \frac{4}{3}$ _____
5. $7 \times \frac{1}{4}$ _____
6. $\frac{3}{4} \times \frac{1}{2}$ _____
7. $\frac{7}{4} \times \frac{6}{2}$ _____
8. $1 \times \frac{1}{5}$ _____
9. $1 \times \frac{5}{3}$ _____
10. $\frac{9}{10} \times 2$ _____

Determine if the answer will be more or less than the original (1st) fraction in the problem. Write <u>less</u> or <u>more</u> for your answers. (DOK 2)

11. $\frac{2}{7} \times \frac{5}{4}$ _____
12. $\frac{3}{4} \times 6$ _____
13. $\frac{2}{3} \times \frac{1}{2}$ _____
14. $\frac{5}{6} \times \frac{3}{4}$ _____
15. $9 \times \frac{3}{2}$ _____

Chapter 6 Multiplying Fractions

6.6 Multiplication Word Problems (DOK 2, 3)

Solve the multi-step problems and simplify answers to lowest terms. Show your work.

1. Sara buys $2\frac{1}{8}$ pounds of grapes every week for 12 weeks. How many pounds of grapes did Sara buy over the 12 week period? _____

2. Beth has a bread machine that makes a loaf of bread that weighs $1\frac{1}{2}$ pounds. If she makes a loaf of bread for each of her three sisters, how many pounds of bread will she make? _____

3. Rick chews on a piece of bubble gum and blows 80 bubbles. He can blow a bubble about every $1\frac{1}{4}$ minutes. How long did it take Rick to blow 80 bubbles? _____

4. Juan can run an average of $1\frac{2}{5}$ feet per second. How many feet can Juan run in 50 seconds? _____

5. Thaddeus needs to measure a length for his garden. He knows his foot measures $1\frac{1}{8}$ feet long. He walks off $13\frac{1}{3}$ foot lengths. How long is his garden? _____

6. Veronica takes $20\frac{1}{2}$ minutes to make one kind of quilt block. There are 12 quilt blocks in the quilt of this kind. It takes her $14\frac{1}{4}$ minutes to make another kind of quilt block. There are 36 quilt blocks of this kind of quilt. How long will it take her to make the 48 quilt blocks? _____

7. Guiseppe timed out that he can do 3 fraction problems in $2\frac{2}{3}$ minutes. His teacher assigned 24 problems for his math homework. How long will it take Guiseppe to do 24 fraction problems? _____

8. Grant has sold $\frac{3}{4}$ of a box of wrapping paper for his school band fund raiser. His mother says she will buy $\frac{1}{3}$ of all the paper left in the box. What fraction of the entire box does Grant's mother buy? _____

9. Shenika found that she can walk one way to school in $7\frac{3}{5}$ minutes. How many minutes does she walk in one week to school? She walks both from home to school and from school to home each day. _____

10. Darius serves $\frac{1}{4}$ pound hamburgers to 27 people at his family reunion. He also serves 58 hotdogs, each weighing $\frac{1}{8}$ lb. How many pounds of meat did Darius serve in all? _____

11. Mr. Barton spends 1 hour per day walking $4\frac{3}{8}$ miles for exercise. How many miles does Mr. Barton walk in 4 days? _____

12. Travis enjoys playing fetch with his dog, Charger. Charger can catch a tennis ball in about $3\frac{3}{10}$ seconds on average. If Travis throws the ball to Charger 75 times every day after school, how many seconds do they play catch each day? _____

13. Marissa is baking for the school carnival. She bakes 14 dozen blueberry muffins, one dozen at a time, for $15\frac{1}{2}$ minutes each dozen. She bakes 6 dozen banana muffins, one dozen at a time, for $14\frac{3}{4}$ minutes each dozen. How long will Marissa spend baking muffins? _____

6.7 Modeling Multiplication

Look at the models in each problem and solve. You may need to do 2–3 steps for each problem. (DOK 3)

1. A shoe store has a rack of 12 pairs of sandals. Isabelle buys $\frac{1}{6}$ of the sandals on display. Twin sisters, Joanie and JoAnn, together buy $\frac{2}{5}$ of the sandals left on the display rack, after Isabelle buys her sandals. How many pairs of sandals are left to sell after Isabelle, Joanie, and JoAnn make their purchases?

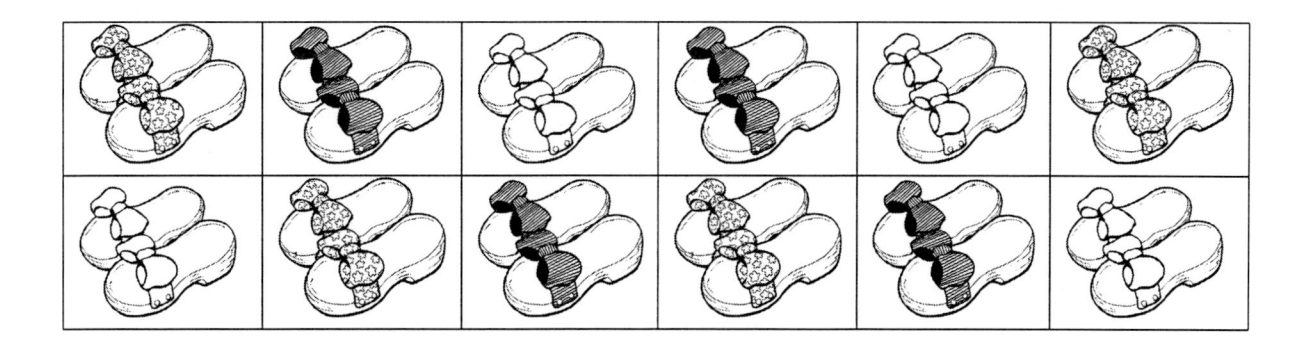

2. Lianna is making bracelets with snap apart pieces. Each bag of pieces has ten parts. Lianna will use $\frac{4}{5}$ of the parts from each bag for each of the 7 bracelets she is making. How many bags of ten parts will she use to make the 7 bracelets?

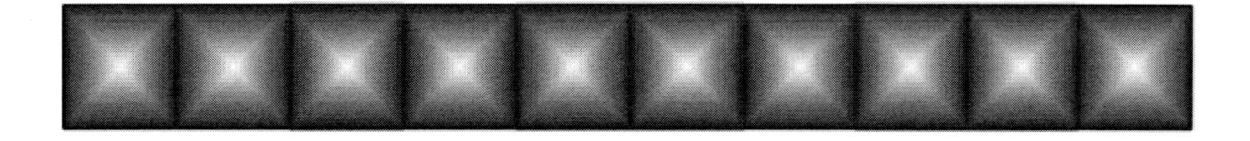

3. The Davis family of six went out to a seafood restaurant, and everyone ordered a plate of 12 shrimp. All 6 of the Davis family ate $\frac{3}{4}$ of their shrimp and took $\frac{1}{4}$ of the shrimp home. How many platefuls of 12 shrimp could they make from the shrimp they took home?

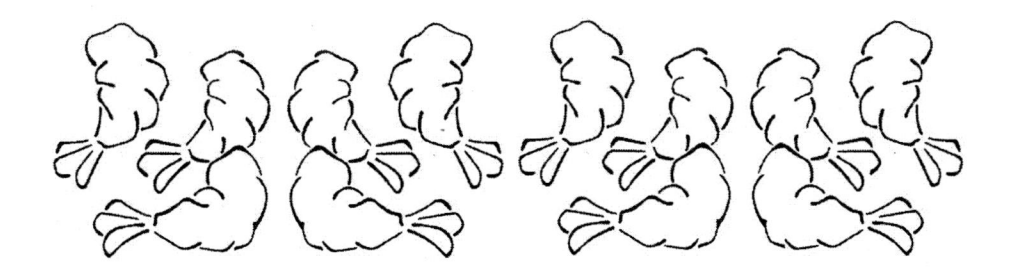

Chapter 6   Multiplying Fractions

4. Mario and his friend, Jacob, counted the number of jelly beans in a bag. They counted how many were licorice favored, the different kinds of fruit flavored, and the coconut flavored jelly beans. They found that in a bag of 66 jelly beans, $\frac{1}{3}$ of them were licorice flavored. How many jelly beans are licorice flavored in one bag? How many are licorice flavored in 4 bags?

5. There are 4 people eating dinner and each ate $\frac{1}{6}$ of the ears of corn pictured below. How many ears of corn did each person eat?

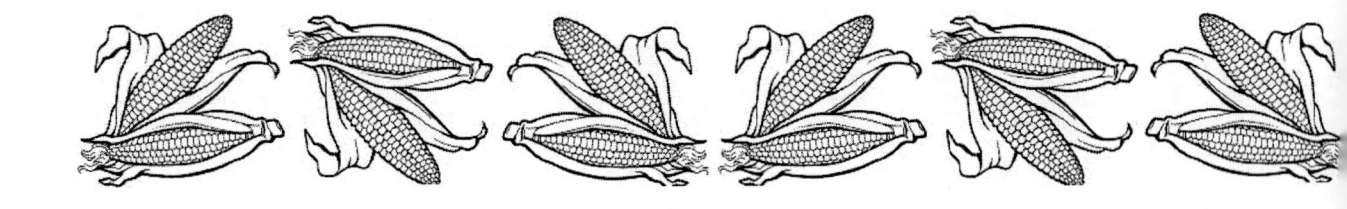

6. Three squirrels each buried $\frac{1}{3}$ of the acorns shown in the picture to the right. Multiply $\frac{1}{3}$ by the number of acorns and find how many acorns each squirrel buried.

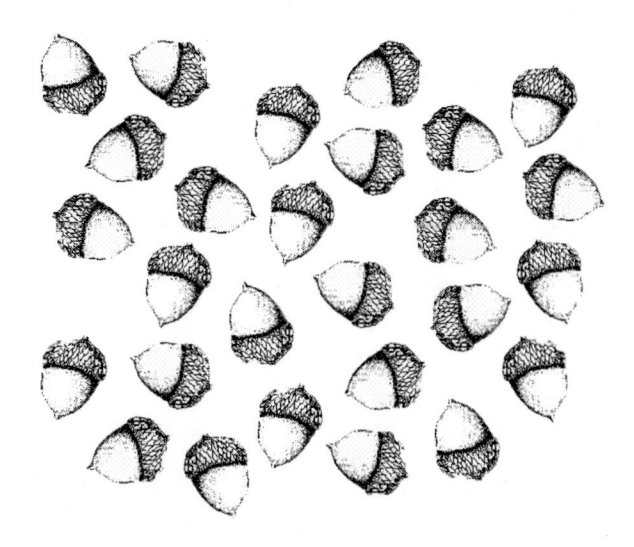

Chapter 6 Review

Multiply the whole numbers by the fractions. Simplify the answers if needed. (DOK 2)

1. $6 \times \frac{2}{3}$ _____
2. $\frac{7}{8} \times 2$ _____
3. $13 \times \frac{2}{5}$ _____
4. $\frac{3}{8} \times 5$ _____

Match the problems in the left column to the model that matches in the right column. (DOK 2)

5. $3 \times \frac{1}{4}$

6. $4 \times \frac{7}{32}$

7. $2 \times \frac{5}{16}$

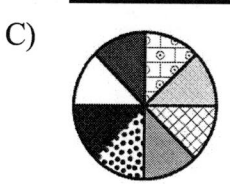

A) | $\frac{1}{4}$ | $\frac{1}{4}$ | $\frac{1}{4}$ | $\frac{1}{4}$ |

B) | $\frac{1}{8}$ | $\frac{1}{8}$ | $\frac{1}{8}$ | $\frac{1}{8}$ | $\frac{1}{8}$ | $\frac{1}{8}$ | $\frac{1}{8}$ | $\frac{1}{8}$ |

C)

Cancel where possible in the following problems, then multiply. (DOK 2)

8. $\frac{2}{5} \times \frac{5}{16}$ _____
9. $\frac{1}{4} \times \frac{4}{7}$ _____
10. $\frac{2}{15} \times \frac{5}{10}$ _____
11. $\frac{3}{11} \times \frac{22}{30}$ _____

Multiply and simplify your answers to lowest terms. (DOK 2)

12. $2\frac{1}{4} \times 3\frac{1}{2}$ _____
13. $\frac{3}{8} \times 5\frac{1}{4}$ _____
14. $1\frac{1}{3} \times 3\frac{1}{6}$ _____
15. $2\frac{2}{5} \times 2\frac{1}{4}$ _____

Look carefully at each problem below. Determine if the answer will be less than or more than one. Write <u>less</u> or <u>more</u> for your answers. (DOK 2)

16. $\frac{5}{6} \times 4$ _____
17. $\frac{1}{4} \times \frac{4}{5}$ _____
18. $\frac{3}{7} \times \frac{9}{8}$ _____
19. $3 \times \frac{3}{2}$ _____

Chapter 6   Multiplying Fractions

Solve and reduce answers to lowest terms. (DOK 2, 3)

20. Mr. Enridge buys $3\frac{1}{5}$ pounds of broccoli every week for 6 weeks. How many pounds of broccoli did Mr. Enridge buy over the 6 week period? _____

21. Sara-Beth has a blender that makes $\frac{1}{4}$ of a gallon of fruit juice. If she makes 3 blender fulls of fruit juice for a picnic party, how many gallons will she make? _____

22. It takes Andrew $\frac{1}{5}$ of a minute to insert one screw into the metal shelving unit he is putting together. The shelving unit has 48 screws in all. How long will it take Andrew to put the unit together? _____

Look at the pizza delivery man model to answer each problem and solve. You may need to do 2–3 steps for each problem. (DOK 3)

23. The pizza delivery man is delivering to an office building. Suite A ordered $\frac{1}{3}$ of the pizzas. Suite B ordered $\frac{1}{4}$ of the pizzas. And Suite C ordered the remainder of the pizzas. How many pizzas does each suite receive?

24. The same pizza delivery man delivers another bundle of pizzas of the same size to a school. Mr. Rickman's 5th grade class receives $\frac{1}{6}$ of the pizzas. Mrs. Bell's 5th grade class receives $\frac{3}{12}$ of the pizzas. The 5th grade class of Mrs. Johnson receives $\frac{1}{2}$ of the pizzas. The rest of the pizzas go to the office ladies. How many pizzas does each class and the office ladies get?

Chapter 6 Test

1 Multiply: $7 \times \frac{1}{2}$. Simplify the answer if needed.

A $7\frac{1}{2}$

B $3\frac{1}{2}$

C 14

D $14\frac{1}{2}$

(DOK 2)

2 Multiply: $11 \times \frac{2}{3}$. Simplify the answer if needed.

A $7\frac{1}{3}$

B $7\frac{2}{3}$

C $22\frac{1}{3}$

D $22\frac{2}{3}$

(DOK 2)

3 Which fraction problem has the answer modeled below?

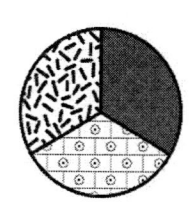

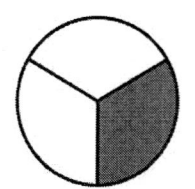

A $1 \times \frac{2}{3}$

B $4 \times \frac{2}{3}$

C $5 \times \frac{1}{3}$

D $4 \times \frac{1}{3}$

(DOK 2)

4 Which fraction problem has the answer modeled below?

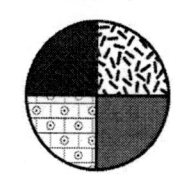

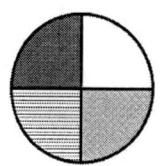

A $1 \times \frac{3}{4}$

B $4 \times \frac{1}{4}$

C $7 \times \frac{1}{4}$

D $8 \times \frac{3}{4}$

(DOK 2)

5 Cancel where possible in the following problem, then multiply.

$$\frac{3}{5} \times \frac{10}{12}$$

A $\frac{1}{2}$

B $\frac{1}{3}$

C $\frac{1}{6}$

D $\frac{1}{5}$

(DOK 2)

6 Multiply: $7\frac{1}{6} \times 2\frac{2}{3}$.

A $14\frac{1}{9}$

B $14\frac{1}{3}$

C $18\frac{5}{6}$

D $19\frac{1}{9}$

(DOK 2)

Chapter 6 Multiplying Fractions

7 Cancel where possible in the following problem, then multiply.

$\dfrac{6}{8} \times \dfrac{4}{12}$

A $\dfrac{1}{2}$

B $\dfrac{1}{3}$

C $\dfrac{1}{4}$

D $\dfrac{1}{5}$

(DOK 2)

8 Multiply: $4\tfrac{1}{4} \times 3\tfrac{1}{3}$.

A $7\tfrac{1}{12}$

B $12\tfrac{1}{12}$

C $14\tfrac{1}{6}$

D $14\tfrac{3}{4}$

(DOK 2)

9 Interpret an answer for $3 \times \tfrac{1}{2}$.

A The answer is less than one.

B The answer is one.

C The answer is more than one.

D The answer is more than two.

(DOK 2)

10 Interpret an answer for $\tfrac{3}{4} \times \tfrac{1}{2}$.

A The answer is less than one.

B The answer is one.

C The answer is more than one.

D The answer is more than two.

(DOK 2)

11 Faye took $\tfrac{1}{2}$ of the package of gum. Her friend, Beth, took $\tfrac{1}{3}$ of the remainder of the package. How much of the package remains?

A $\dfrac{1}{2}$

B $\dfrac{1}{3}$

C $\dfrac{1}{5}$

D $\dfrac{1}{6}$

(DOK 3)

12 Joshua has 5 pieces of wood that each measure $3\tfrac{3}{4}$ feet long. He uses $\tfrac{1}{2}$ of the wood for a school project. He uses $\tfrac{1}{4}$ of the wood to make a bird house for his mother. How much wood does Joshua have left over?

A $4\tfrac{5}{16}$ feet

B $4\tfrac{1}{4}$ feet

C $4\tfrac{11}{16}$ feet

D $4\tfrac{1}{8}$ feet

(DOK 3)

13 Tim was asked what fraction of the grid below is ♣ ?

A $\dfrac{1}{2}$

B $\dfrac{2}{3}$

C $\dfrac{5}{12}$

D $\dfrac{5}{6}$

(DOK 2)

90

Chapter 7
Dividing Fractions

This chapter covers the following CC 5 standards:

| Number and Operations - Fractions | 5.NF.3, 5.NF.7 |

7.1 Interpreting Fractions as Division Problems (DOK 2)

A fraction is actually a division problem.

Example 1: The fraction $\frac{1}{4}$ is the same as the division problem: $1 \div 4$.

$1 \div 4 = \frac{1}{4}$

One pie divided into 4 pieces is shown below. $1 \div 4$.

Each piece is equal to $\frac{1}{4}$ of a pie. $\dfrac{1 \text{ piece}}{4 \text{ pieces}}$

Example 2: There are 7 pounds of apples to be divided equally by 5 people. What portion does each person get?

Step 1: Set up a division problem. 7 pounds of apples to share with 5 people: $7 \div 5$

Step 2: Make the division problem into a fraction: $7 \div 5 = \frac{7}{5}$, simplified $= 1\frac{2}{5}$ pounds per person.

Step 3: Multiply to check your answer: $\left(1\frac{2}{5} \times 5\right) = \left(\frac{7}{5} \times 5\right) = \frac{35}{7} = 7$

Answer: You are correct - Each person receives $1\frac{2}{5}$ pounds of apples.

Chapter 7   Dividing Fractions

Set up each division problem as a fraction and solve. Give your answers in simplest form. (DOK 2)

1. Jerry's pack of gum has 14 sticks. He shares the pack of gum with his sister and brother. What portion of the pack of gum will the 3 children get if they all get an equal amount? Between what two whole numbers does the answer lie? _____ ; _____

2. Mrs. Hanson brought home a pizza divided into 8 equal pieces. There are 3 people in the family. If each person gets the same amount, what portion of the pizza will each person get? _____

3. Darlene has 16 yards of yarn. She has 3 cats and is making cat toys with the yarn. She uses the same amount of yarn to make 3 cat toys, one for each cat. What portion of the yarn is used in each cat toy? Between what two whole numbers does the answer lie? _____ ; _____

4. James bought a loaf of bread with 20 slices in it. He makes sandwiches using the whole loaf for 10 people. What portion of the loaf does each person receive? _____

5. A package of Goody Graham Crackers contains 40 graham crackers. Six friends divide the package equally between themselves. How many graham crackers does each friend get? Between what two whole numbers does the fraction lie? _____ ; _____

6. There were 11 cookies left in the bag. Three friends shared the cookies. What portion of the cookies did each friend receive? _____

7. Seventeen of the monkeys at the zoo like bananas. There are 51 monkeys in all at the zoo. What fraction of the monkeys like bananas? Between what two whole numbers does the answer lie? _____ ; _____

8. The six Tritts went on a family car trip with 20 sandwiches. If each member of the Tritt family received the same amount of sandwiches, how many sandwiches did each Tritt eat? _____

9. Mrs. Smith bought a bag of oranges containing 17 oranges. Each of the five members of the Smith family shared the oranges equally. What portion did each Smith get? Between what two whole numbers does the fraction lie? _____ ; _____

10. Elisa has 31 pieces of paper left. She shares the paper equally between herself and her friend Beatrice. What portion of the paper does each person get? Between what two whole numbers does the answer lie? _____ ; _____

11. A mother bird has 4 nestlings to feed. Over a one hour period one morning, she brings 6 bugs to the nest. If each nestling got an equal share, what portion of the bugs did each nestling receive? _____

12. Veronica has to read a book that is 123 pages long, in four evenings. If she reads an equal number of pages each night, what portion of the book must Veronica read each night? _____

7.2 Reciprocals (DOK 1)

The **reciprocal** of a fraction is found by switching the numerator and the denominator.

Example 3: Find the reciprocal of the fraction $\frac{2}{3}$.

  Step 1: Switch the numerator and denominator. $\frac{3}{2}$

  Answer: The reciprocal of the fraction $\frac{2}{3}$ is $\frac{3}{2}$.

The reciprocal is the fraction turned upside down. The reciprocal of $\frac{5}{6}$ is $\frac{6}{5}$. The reciprocal of $\frac{1}{3}$ is $\frac{3}{1}$. The reciprocal of $\frac{742}{933}$ is $\frac{933}{742}$.

Find the reciprocal of each fraction. You do <u>not</u> need to simplify the fractions. (DOK 1)

1. $\frac{1}{2}$ _____
2. $\frac{8}{11}$ _____
3. $\frac{13}{17}$ _____
4. $\frac{2}{5}$ _____
5. $\frac{3}{9}$ _____

6. $\frac{4}{10}$ _____
7. $\frac{12}{15}$ _____
8. $\frac{3}{7}$ _____
9. $\frac{2}{9}$ _____
10. $\frac{1}{11}$ _____

11. $\frac{10}{40}$ _____
12. $\frac{3}{8}$ _____
13. $\frac{11}{13}$ _____
14. $\frac{4}{5}$ _____
15. $\frac{1}{7}$ _____

16. $\frac{3}{5}$ _____
17. $\frac{10}{12}$ _____
18. $\frac{2}{6}$ _____
19. $\frac{1}{8}$ _____
20. $\frac{4}{6}$ _____

Chapter 7 Dividing Fractions

7.3 Dividing Fractions by Whole Numbers (DOK 2)

Example 4: Brandon divided a half of a piece of paper into four parts. $\frac{1}{2} \div 4 = \frac{1}{8}$

Each part of the half piece is equal to $\frac{1}{8}$ of a piece of paper.

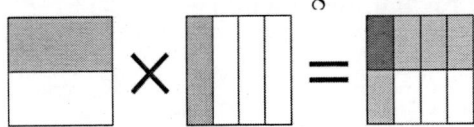

Brandon could have figured how much each part of the piece of paper would be mathematically. $\frac{1}{2} \div 4$

Step 1: Change 4 to a fraction: $\frac{4}{1}$.
Step 2: Find the reciprocal of the second fraction: $\frac{1}{4}$
Step 3: Multiply the two fractions: $\frac{1}{2} \times \frac{1}{4} = \frac{1}{8}$
To check: $\frac{1}{8} \times 4 = \frac{4}{8} = \frac{1}{2}$

Example 5: Alissa took $\frac{3}{4}$ of a box of cereal and divided it into 6 equal parts. What fraction is each part of the box of cereal?

Solve: $\frac{3}{4} \div 6 = \frac{3}{4} \div \frac{6}{1} = \frac{3}{4} \times \frac{1}{6} = \frac{3}{24}$, simplified $= \frac{1}{8}$
To check: $\frac{1}{8} \times 6 = \frac{6}{8} = \frac{3}{4}$

Give a two part answer: First, divide the fractions by the whole numbers. Simplify if needed. Then multiply your quotient by the whole number to double check your answer. The first one is done for you. (DOK 2)

1. $\frac{7}{8} \div 2 = \frac{7}{16}$
   $\frac{7}{16} \times 2 = \frac{14}{16} = \frac{7}{8}$

2. $\frac{2}{3} \div 6$ _____

3. $\frac{9}{10} \div 3$ _____

4. $\frac{4}{5} \div 4$ _____

5. $\frac{3}{5} \div 2$ _____

6. $\frac{6}{9} \div 3$ _____

7. $\frac{2}{7} \div 3$ _____

8. $\frac{6}{8} \div 4$ _____

9. $\frac{1}{8} \div 2$ _____

10. $\frac{2}{9} \div 3$ _____

11. $\frac{3}{4} \div 6$ _____

12. $\frac{3}{8} \div 2$ _____

13. $\frac{2}{10} \div 4$ _____

14. $\frac{9}{12} \div 3$ _____

15. $\frac{3}{6} \div 4$ _____

16. $\frac{7}{9} \div 2$ _____

7.4 Dividing Whole Numbers by Fractions (DOK 2)

Example 6: Majella took 4 pears and divided them into parts equaling $\frac{1}{2}$ each. How many parts of the four pears does Majella have? $4 \div \frac{1}{2}$

Step 1: Turn the whole number into a fraction: $4 = \frac{4}{1}$.

Step 2: Multiply the fraction form of the whole number by the **reciprocal** of the fraction in the problem: $\frac{4}{1} \times \frac{2}{1} = 8$. Majella now has 8 parts of pears.

To check: Multiply the quotient by the fraction in the original problem and simplify: $8 \times \frac{1}{2} = \frac{8}{2} = 4$. It is correct.

Refer to the "About AR" on page xi!

Example 7: Solve $4 \div \frac{1}{5}$

Step 1: Turn the whole number into a fraction: $4 = \frac{4}{1}$.

Step 2: Find the reciprocal of $\frac{1}{5} = \frac{5}{1}$.

Step 3: Multiply: $\frac{4}{1} \times \frac{5}{1} = \frac{20}{1} = 20$

To check: $\frac{20}{1} \times \frac{1}{5} = \frac{20}{5} = 4$

Divide the whole numbers by fractions using the method above. Give a two part answer: First, divide the whole numbers by the fractions. Simplify if needed, but leave your answers as improper fractions. Then multiply your quotient by the fraction in the original problem to double check your answer. The first one is done for you. (DOK 2)

1. $3 \div \frac{4}{5} =$

$\frac{3}{1} \times \frac{5}{4} = \frac{15}{4}$

$\frac{15}{4} \times \frac{4}{5} = \frac{60}{20} = 3$

2. $5 \div \frac{1}{2} =$ _____

3. $7 \div \frac{2}{3} =$ _____

4. $4 \div \frac{1}{3} =$ _____

5. $8 \div \frac{2}{5} =$ _____

6. $2 \div \frac{3}{4} =$ _____

7. $10 \div \frac{2}{7} =$ _____

8. $6 \div \frac{3}{7} =$ _____

9. $11 \div \frac{1}{2} =$ _____

10. $1 \div \frac{3}{8} =$ _____

11. $3 \div \frac{2}{9} =$ _____

12. $12 \div \frac{5}{6} =$ _____

13. $6 \div \frac{2}{3} =$ _____

14. $5 \div \frac{1}{4} =$ _____

15. $16 \div \frac{1}{2} =$ _____

Chapter 7 Dividing Fractions

7.5 Division Word Problems (DOK 2)

Solve and simplify answers to lowest terms. (DOK 2)

1. How many $\frac{3}{4}$ cups servings are in 4 cups of milk? Hint: $4 \div \frac{3}{4}$. _____

2. If 3 dogs equally share a $\frac{1}{2}$ pound of dog food, what share will each dog get? _____

3. Jennifer and four of her classmates are sharing $\frac{3}{4}$ of a package of ink pens. What fraction of the package of pens will each classmate receive? _____

4. Bryan, Ryan, and Jack will equally share 2 apples. What fraction of an apple will each boy receive? _____

5. Mr. Tritt shares $\frac{1}{2}$ bag of potting soil equally between 4 flower pots. What fraction of a bag of potting soil will each flower pot get? _____

6. Mr. and Mrs. Bowman are equally sharing the last $\frac{3}{8}$ of a pizza, left over from last night. What fraction of a pizza will each of the couple get? _____

7. There are 7 people at a birthday party. They will equally share $\frac{3}{4}$ of the birthday cake. What fraction of the birthday cake will each person receive? _____

8. Marlisse and her sister, Alexis, are equally sharing $\frac{1}{2}$ of a package of beads to make bracelets. What fraction of the package of beads will each girl get? _____

9. Three brothers are equally sharing the last $\frac{4}{5}$ of a box of cereal. What fraction of the box of cereal will each boy receive? _____

10. There is $\frac{1}{2}$ of a bucket of oats left in the barn. Farmer John will give each of his 4 horses an equal share of the last of the oats, while waiting for delivery of more oats. What fraction of a bucket of oats will each horse receive? _____

11. Jason decides to divide the last two hours before bedtime equally between completing his homework, cleaning his room, and playing a video game. What fraction of the last 2 hours of the day can Jason devote to each item? _____

12. Lucy's grandmother has 5 grandchildren over for a visit. She wants to equally divide $\frac{7}{8}$ of a bag of butterscotch candies among the children. What fraction of a bag of butterscotch candies will each grandchild receive? _____

13. Alisha has $\frac{2}{3}$ of a box of cat treats left. She has 4 cats. If Alisha gives each cat an equal share of the cat treats, what fraction of the box will each cat get? _____

14. How many $\frac{1}{2}$ cup servings are in 3 cups of cole slaw? _____

7.6 Modeling Division (DOK 2, 3)

Look at the models in each problem and solve. You may need to do 2–3 steps for each problem. (DOK 2, 3)

1. Aaron folded a piece of paper into 2 parts. He drew grid lines on one part of the paper and shaded in $\frac{1}{2}$ of the gridded side. What fraction of the whole piece of paper has both grid lines and shading?

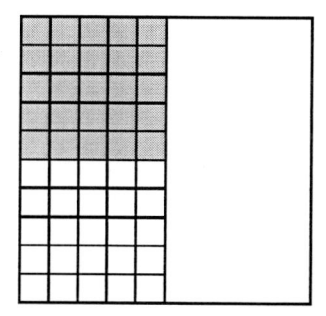

2. Traci had 6 pieces of taffy. She ate $\frac{1}{3}$ of the pieces. She then gave $\frac{1}{2}$ of the remaining pieces to her friend, Annette. What fraction of the six pieces did Annette receive?

3. Jonathan is trying to play marbles. He takes 15 marbles out of his jar. His kitten named Frisky, bats 5 of the marbles Jonathan is playing with, under the bed. What fraction of the marbles did Frisky bat under the bed?

Chapter 7 Dividing Fractions

4. Roger took a piece of cardboard and divided it into 3 pieces.

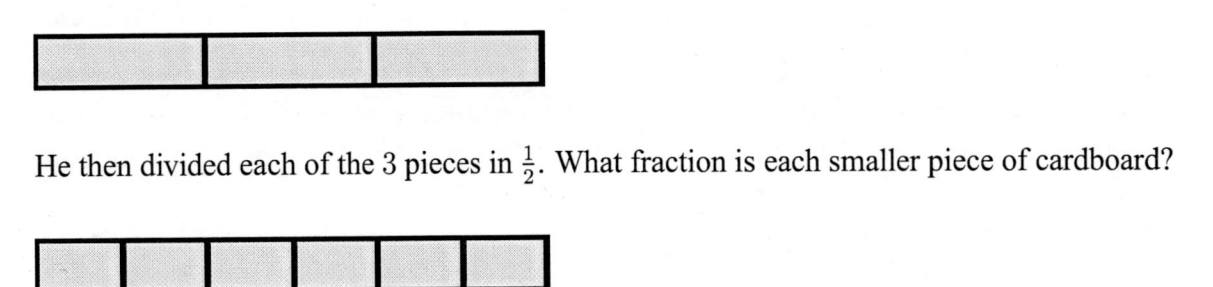

He then divided each of the 3 pieces in $\frac{1}{2}$. What fraction is each smaller piece of cardboard?

5. Tanya folded a bath towel in $\frac{1}{2}$, then folded it again in $\frac{1}{2}$. She then made one more fold in $\frac{1}{2}$ again.

If you are looking at the top of this multi-folded towel, what fraction of the towel can you see?

6. Sally Ann picked 16 strawberries. She divided the strawberries and gave $\frac{1}{4}$ to her brother. She kept the rest for herself. What fraction of the strawberries did Sally Ann keep for herself?

Chapter 7 Review

Set up each division problem as a fraction and solve. Simplify your answers. (DOK 2)

1. Gail's pack of gum has 15 sticks. She shares the pack of gum equally with herself, her sister, and a friend. What fraction of the pack of gum will each person get? _____

2. Mr. Tillman brought home a loaf of bread divided into 20 slices for his family of 4. If each person gets the same amount, what fraction of the the slices of bread will each person get? _____

Give a two part answer: First, divide the fractions by the whole numbers. Simplify if needed. Then multiply your quotient by the whole number in the problem to double check your answer. (DOK 2)

3. $\frac{3}{5} \div 2$ _____   4. $\frac{2}{3} \div 4$ _____   5. $\frac{9}{15} \div 3$ _____   6. $\frac{6}{7} \div 2$ _____

Find the reciprocal of each fraction. You do <u>not</u> need to simplify the fractions. (DOK 1)

7. $\frac{3}{5}$ _____   8. $\frac{2}{13}$ _____   9. $\frac{21}{23}$ _____   10. $\frac{1}{4}$ _____

Divide the whole numbers by fractions using reciprocals. Give a two part answer: First, divide the whole numbers by the fractions. Simplify if needed. Then multiply your quotient by the whole number to double check your answer. (DOK 2)

11. $8 \div \frac{2}{5}$ _____   12. $4 \div \frac{2}{3}$ _____   13. $10 \div \frac{1}{4}$ _____   14. $6 \div \frac{1}{3}$ _____

Solve and simplify answers to lowest terms. (DOK 2)

15. How many $\frac{2}{3}$ cups servings are in 5 cups of water? _____

16. If 6 gerbils equally share a $\frac{1}{3}$ pound of gerbil food, what share will each gerbil get? _____

Look at the models in each problem and solve. You may need to do 2–3 steps for each problem. (DOK 2)

17. Trevor folded a piece of paper into 3 parts, vertically. He then folded it again into $\frac{1}{3}$, this time horizontally. Trevor then unfolded the piece of paper to see the many folds. What fraction is one piece of the paper without any folds?

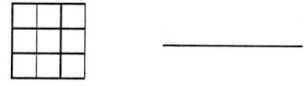

18. Anya has a book of bird theme stickers. One page of 12 stickers is all doves. She stuck $\frac{1}{4}$ of the dove stickers below on her English notebook.

    How many dove stickers did Anya stick on her English notebook? _____

Chapter 7 Dividing Fractions

Chapter 7 Test

1 A farmer hauled in 120 bales of hay. Each of his cows ate $\frac{1}{2}$ of a bale. How many cows did the farmer feed?

A $120 \div \frac{1}{2} = \frac{240}{2} = 120$

B $120 \div \frac{1}{2} = \frac{240}{1} = 240$

C $120 \div \frac{1}{2} = \frac{120}{2} = 60$

D $120 \div \frac{1}{2} = \frac{120}{1} = 120$

(DOK 2)

2 Martin makes sandwiches for his family. He has 32 ounces of sandwich meat. If he makes 12 sandwiches, how many ounces of meat did Martin put in each sandwich?

A $2\frac{1}{2}$

B $2\frac{2}{3}$

C $2\frac{1}{3}$

D $2\frac{1}{18}$

(DOK 2)

3 Divide: $\frac{3}{4} \div 6$

A $\frac{1}{8}$

B $4\frac{1}{2}$

C $\frac{1}{4}$

D $\frac{1}{2}$

(DOK 2)

4 Which fraction is the reciprocal of $\frac{5}{11}$?

A 5

B 11

C $\frac{5}{11}$

D $\frac{11}{5}$

(DOK 1)

5 Divide: $7 \div \frac{1}{2}$

A $\frac{1}{14}$

B $\frac{7}{2}$

C 14

D $\frac{2}{7}$

(DOK 2)

Chapter 7 Test

6 A box of 10 fruit flavored ice treats came with $\frac{1}{5}$ cherry flavored. If the box is divided into 2 parts with the same flavors in each part, what fraction of each part is cherry flavored? Express your answer in simplest form.

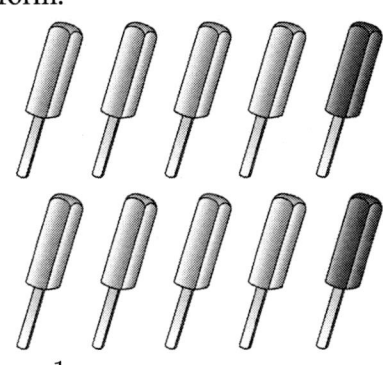

A $\frac{1}{10}$

B $\frac{4}{5}$

C $\frac{1}{5}$

D $\frac{2}{5}$

(DOK 2)

7 Sven has $\frac{1}{2}$ box of crackers. If he divides the crackers so he and his 2 brothers receive an equal amount, what fraction of the entire box will each of the brothers receive?

A $\frac{1}{2}$

B $\frac{1}{3}$

C $\frac{1}{6}$

D $\frac{1}{12}$

(DOK 2)

8 Divide: $9 \div \frac{2}{3}$

A $13\frac{1}{2}$

B $9\frac{2}{3}$

C $\frac{1}{18}$

D $\frac{2}{27}$

(DOK 2)

9 Zeke has $\frac{5}{6}$ of a package of cookies to share with his 4 friends and himself. One of his friends keeps $\frac{1}{2}$ of his share to take home. How much of the partial package did Zeke's friend take home?

A $\frac{5}{6}$

B $\frac{1}{2}$

C $\frac{1}{6}$

D $\frac{1}{12}$

(DOK 2)

10 There are $\frac{7}{10}$ of a pound of grapes left. Three sisters decide to divide the grapes evenly among them, each receiving $\frac{1}{3}$ of the grapes. Lydia gives $\frac{1}{4}$ of her share to her friend, Sandy. What fraction of a pound of grapes did Sandy receive?

A $\frac{7}{120}$

B $\frac{1}{40}$

C $\frac{7}{40}$

D $\frac{1}{10}$

(DOK 2)

Copyright © American Book Company

Chapter 8
Measurements and Line Plots

This chapter covers the following CC 5 standards:

Measurement and Data	5.MD.1, 5.MD.2

8.1 Customary Measurement (DOK 1)

Customary measure in the United States is based on the English system. The following chart gives common customary units of measure as well as the standard units for time.

English System of Measure

Measure	Abbreviations	Appropriate Instrument
Time: 1 week = 7 days 1 day = 24 hours 1 hour = 60 minutes 1 minute = 60 seconds	week = wk hour = hr or h minutes = min seconds = sec	calendar clock clock clock
Length: 1 mile = 5,280 feet 1 yard = 3 feet 1 foot = 12 inches	mile = mi yard = yd foot = ft inch = in	odometer yard stick, tape line ruler, yard stick
Volume: 1 gallon = 4 quarts 1 quart = 2 pints 1 pint = 2 cups 1 cup = 8 ounces	gallon = gal quart = qt pint = pt ounce = oz	quart or gallon container quart container cup, pint, or quart container cup
Weight: 1 pound = 16 ounces 2,000 pounds = 1 ton	ounce = oz pound = lb Ton = T	scale or balance scale or balance scale
Temperature: Fahrenheit Celsius	°F °C	thermometer thermometer

8.2 Converting Units of Customary Measure (DOK 2)

Equivalents:

1 mile	=	1,760 yards	=	5,280 feet
1 yard	=	3 feet	=	36 inches
1 foot	=	12 inches		

1 pound	=	16 ounces

1 gallon	=	4 quarts	=	8 pints
1 quart	=	2 pints	=	4 cups
1 pint	=	2 cups	=	16 ounces
1 cup	=	8 ounces		

Using the information above, fill in the blanks below. (DOK 2)

1. 32 ounces = _____ pound(s)
2. 36 inches = _____ yard(s)
3. 3 yards = _____ feet
4. 2 gallons = _____ quart(s)
5. 5 pints = _____ cup(s)
6. $1\frac{1}{2}$ gallons = _____ quart(s)
7. $1\frac{1}{2}$ pints = _____ ounce(s)
8. 2 yards = _____ inches
9. 60 inches = _____ feet

10. 2 quarts = _____ gallon(s)
11. 1 quart = _____ ounces
12. $\frac{3}{4}$ quart = _____ cups
13. 3 pints = _____ ounces
14. 18 inches = _____ feet
15. 3 gallons = _____ pint(s)
16. 5 cups = _____ pint(s)
17. 5 quarts = _____ gallon(s)
18. 4 feet = _____ yard(s)

19. Mrs. Drake is using a recipe for meatloaf that calls for 8 ounces of sausage. If she triples the recipe for a family reunion party, how many pounds of sausage will she need?

20. Alicia is making ice tea for herself and three friends. If each person drinks 16 ounces of ice tea, how many quarts of ice tea should Alicia make?

21. Marcus needs to buy three gallons of ice cream for his birthday party. The ice cream he wants only comes in cartons equal to 2 quarts. How many cartons of ice cream will he need to buy?

Chapter 8   Measurements and Line Plots

8.3   Real-World Customary Measurement (DOK 2)

Some of the word problems below will require 2–3 steps to solve. (DOK 2)

1. Shannon made 2 quarts of lemonade. She poured 8 ounces of the lemonade into each glass around the dinner table. There was not any lemonade left over. How many 8 ounces glasses did Shannon pour the lemonade into?

2. Mr. O'Malley bought a 12 foot length of lumber to make bird houses. He used 1 yard of the length of lumber for each bird house. How many bird houses did Mr. O'Malley build?

3. Alissa and her mother ran 13,200 feet together on Saturday morning. How many miles did they run?

4. Beth and Anne's father said he would pay the two girls $0.12 an ounce for digging up dandelions out of the yard. Beth earned $3.84 for her pile of dandelions and Anne earned $4.80 for her pile of dandelions. How many pounds of dandelions did Beth and Anne dig out of the yard all together?

5. Mrs. Rosker needed 1 gallon of sliced peaches to make all the peach pies for the family reunion. If each peach pie requires 4 cups of sliced peaches, how many pies will Mrs. Rosker be making?

6. Jack weighs 78 pounds when he stands on the scale holding his 120 ounce Chihuahua, Choppers. How many pounds does Jack weigh?

7. Margaret bought a $2\frac{1}{2}$ yard length of fabric. She used 24 inches of the length to make a doll dress. She used the remainder to make a sundress for herself. What length, in inches, did Margaret use to make her sundress?

8. Andrew rode exactly 1 mile on his bike to the library. He then rode 1,320 feet to meet his friends at the park. Finally, he rode his bike 880 yards to his friend's house. How many feet did Andrew ride his bike in all?

9. Paul divided 4 quarts of milk into 4 pint jars, and the remainder into 8 ounce jars. How many 8 ounce jars did Paul use?

10. Shayla was having a difficult time getting her two cats to stand on the bathroom scale one at a time so she could weigh them. Then she got an idea! She held both cats in her arms and got on the scale herself. With both cats, Shayla weighed 82.6 pounds. One cat jumped from her arms and the scale now said 75.2 pounds. She then set the second cat down and stepped on the scale by herself to find she weighs 67.4 pounds. How much do each of the cats weigh?

11. Jason measured the length of the living room and found it to be 4 yards, 2 feet, and 3 inches long. His father measured the length of the living room and found it to be 168 inches. Did they get the same measure, and if not what is the difference between the two measures?

12. Brandon is building a dog pen that measures 25 feet long and 18 feet wide. How many yards of fencing will he need?

8.4 Metric Measurement (DOK 1)

The metric system uses units based on multiples of ten. The basic units of measure in the metric system are the meter, the liter, and the gram. Metric prefixes tell what multiple of ten the basic unit is multiplied by. Below is a chart of metric prefixes and their values. The ones rarely used are shaded.

Prefix	kilo (k)	hecto (h)	deka (da)	unit (m, L, g)	deci (d)	centi (c)	milli (m)
Meaning	1000	100	10	1	0.1	0.01	0.001

Multiply when changing from a greater unit to a smaller one; **Divide** when changing from a smaller unit to a larger one. **The chart is set up to help you know how far and which direction to move a decimal point when making conversions from one unit to another.**

8.5 Understanding Meters (DOK 1)

The basic unit of **length** in the metric system is the **meter**. Meter is abbreviated "m."

Metric Unit	Abbreviation	Memory Tip	Equivalents
1 millimeter	mm	Thickness of a dime	10 mm = 1 cm
1 centimeter	cm	Width of the tip of the little finger	100 cm = 1 m
1 meter	m	Distance from the nose to the tip of fingers (a little longer than a yard)	1000 m = 1 km
1 kilometer	km	A little more than half a mile	

8.6 Understanding Liters (DOK 1)

The basic unit of **liquid volume** in the metric system is the **liter**. Liter is abbreviated "L." A capital L is used to signify liter, so it is not confused with the number 1. The liter is the volume of a cube measuring 10 cm on each side. Any liquid, such as soda or gasoline, can be measured in liters.

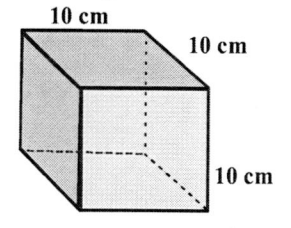

Volume = 1000 cm^3 = 1 liter
(a little more than a quart)

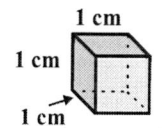

Volume = 1 cm^3 = 1 mL
(an eyedropper holds 1 mL)

8.7 Understanding Grams (DOK 1)

The basic unit of **mass** in the metric system is the **gram**, which is about the mass of a large paperclip. Gram is abbreviated "g."

1000 grams = 1 kilogram (kg) = a little over 2 pounds. People, animals, and objects are weighed in kilograms.

1 milligram (mg) = 0.001 gram. This is an extremely small amount used in medicine. An aspirin tablet is 300 mg.

8.8 Converting Units Within the Metric System (DOK 2)

Converting units such as kilograms to grams or centimeters to decimeters is easy now that you know how to multiply and divide by multiples of ten.

Prefix	kilo (k)	hecto (h)	deka (da)	unit (m, L, g)	deci (d)	centi (c)	milli (m)
Meaning	1000	100	10	1	0.1	0.01	0.001

Example 1: 2 L = ____ mL

2.000 L = 2000 mL

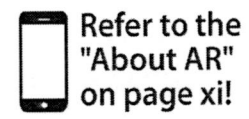

Look at the chart above. To move from liters to milliliters, you move to the right three places. So, to convert the 2 L to mL, move the decimal point three places to the right. You will need to add three zeros.

Refer to the "About AR" on page xi!

Example 2: 5.2 cm = ____ m

005.2 cm = 0.052 m

To move from centimeters to meters, you need to move two spaces to the left. So, to convert 5.2 cm to m, move the decimal point two spaces to the left. Again, you need to add zeros.

Solve the following problems. (DOK 2)

1. 35 mg = ____ g
2. 6 km = ____ m
3. 21 mL = ____ L
4. 4.9 mm = ____ cm
5. 5.35 kL = ____ mL
6. 32 g = ____ kg
7. 156 m = ____ km
8. 25 mg = ____ cg
9. 17.5 L = ____ mL
10. 4 g = ____ kg
11. 0.6 cL = ____ L
12. 0.417 kg = ____ cg
13. 18.2 cL = ____ L
14. 81.2 dm = ____ cm
15. 72.3 cm = ____ m
16. 0.003 kL = ____ L
17. 5.06 g = ____ mg
18. 1.05 mL = ____ cL
19. 43 hm = ____ km
20. 2.057 m = ____ cm
21. 564 g = ____ kg

8.9 Real-World Metric Measurement (DOK 2)

Some of the word problems below will require 2–3 steps to solve. (DOK 2)

1. Marco rode with his father 83.5 kilometers one way to his uncle's house. While there, he walked his uncle's dog 2,250 meters around the neighborhood. Marco and his father drove home after the visit. How many kilometers did Marco ride and walk in all on the trip?

2. Lisa was astonished to learn her dog, Bubbles, weighs 22,500 grams. How many kilograms does Bubbles weigh?

3. Francis bought 1,100 centimeters of rope to make a yard swing. He only used 8 meters of the rope for the swing. How many centimeters of rope did Francis have left over?

4. Brent measured the length of his bed and found it is 220 centimeters long. How many meters long is Brent's bed?

5. Joshua was thirsty and drank 1 liter of water. He then drank another 250 mL of water. Later that day, he drank another 2,000 mL of water. How many liters of water did Joshua drink in the whole day?

6. In science class, the students were asked to weigh their grasshoppers before studying them. Isabelle's grasshopper weighed 25 grams. Her lab partner, Marie, weighed the same grasshopper and found it to be 25,000 milligrams. Did the two girls get the same measure for their grasshopper, and if not, what is the difference in weight?

7. Taylor poured exactly 2 liters of water into 4 glasses, but without getting the volume in each glass even. She then poured each glass of water out and found glass #1 had 450 mL in it, glass #2 had 510 mL in it, and glass #3 had 490 mL in it. What was the volume of water in glass #4?

8. Mrs. Engleson took a package and 2 letters to the post office in a foreign country where they found the package weighed 1.2 kilograms, the first letter weighed 25 grams and the second letter weighed 42 grams. How much did the 3 items weigh in kilograms, all together?

9. Julie walked 0.9 kilometers from home to the grocery store, then 300 meters to her mother's office to bring her lunch. How many kilometers did Julie walk from home to the grocery store, and then to her mother's office?

10. Johan's cousin gave him 300 meters of thin wire to use for a class project. He used 230 centimeters to wire his papier-mache planetary system in a diorama. He used another 180 centimeters to hang a small wind chime on his bedroom window. How much wire, in meters, is there left over?

11. After dinner, Onalisse was asked to pour all the leftover drinks into a liter bucket and pour it down the sink. One glass still had 120 mL of water in it, and another had 80 mL of milk in it. Two glasses each had 0.2 L of water in them. How much liquid did Onalisse pour into the liter bucket?

Chapter 8 Measurements and Line Plots

8.10 Line Plots (DOK 2)

Line plots are used to display a set of measurements.

Example 3: The line plot below shows the number of and weights of fish caught in the Billy's Bass Tournament.

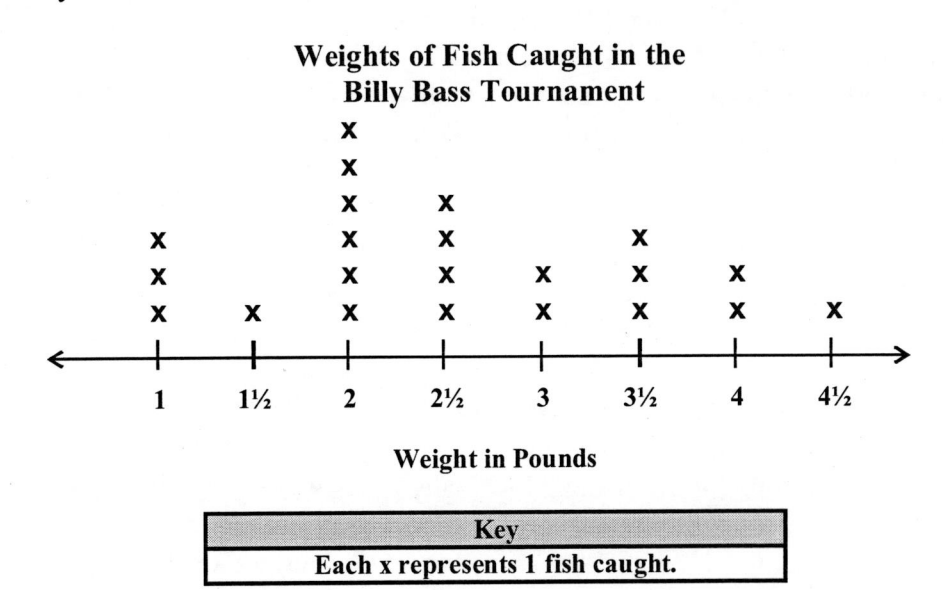

Weights of Fish Caught in the Billy Bass Tournament

```
                    x
                    x
                    x   x
         x          x   x           x
         x          x   x   x   x   x
         x   x      x   x   x   x   x   x
    <----+---+---+---+---+---+---+---+---->
         1  1½   2  2½   3  3½   4  4½
```

Weight in Pounds

Key
Each x represents 1 fish caught.

The weight in pounds of the fish caught are on the horizontal bar at the bottom of the graph.

The number of fish caught are represented by the x's above each weight.

By choosing the weight first, you can move up the graph to see how many fish were caught at that weight, by counting how many x's are plotted.

Question: How many fish were caught weighing 3 pounds? Go to the 3 pound mark along the bottom of the line plot. Then, count how many x's are plotted above the 3 pound mark. There were 2 fish caught weighing 3 pounds.

Question: How many fish were caught weighing $2\frac{1}{2}$ pounds? Go to the $2\frac{1}{2}$ pound mark along the bottom of the line plot. Then, count how many x's are plotted above the $2\frac{1}{2}$ pound mark. There were 4 fish caught weighing $2\frac{1}{2}$ pounds.

8.10 Line Plots (DOK 2)

Example 4: Mrs. Wade made enough lemonade for each of her students to receive 2 cups. Each student was asked to drink the amount of lemonade they wanted for 1 minute. They were then asked to measure the amount of lemonade left in their glass. The results of the leftover lemonade are shown in the line plot below.

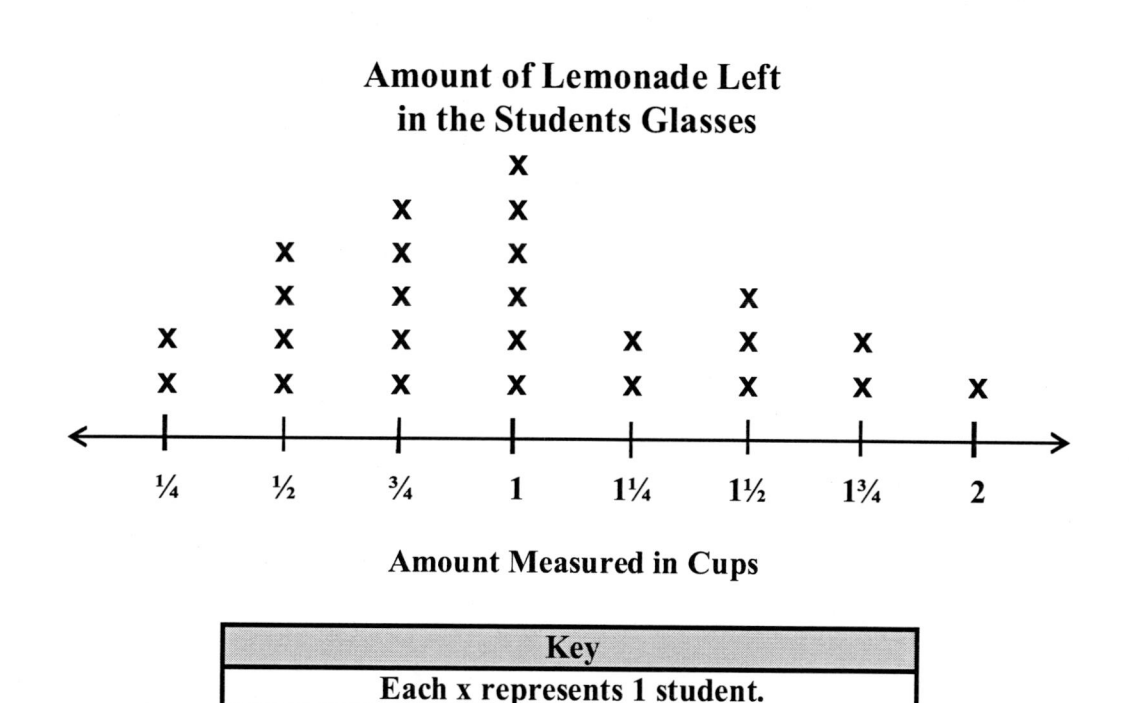

Amount of Lemonade Left in the Students Glasses

```
                        x
                  x     x
            x     x     x
            x     x     x           x
      x     x     x     x     x     x     x
      x     x     x     x     x     x     x     x
   ───┼─────┼─────┼─────┼─────┼─────┼─────┼─────┼───▶
      ¼     ½     ¾     1    1¼    1½    1¾     2
```

Amount Measured in Cups

Key
Each x represents 1 student.

Question: How many students had $\frac{3}{4}$ of a cup of leftover lemonade?
There are 5 x's above the $\frac{3}{4}$ mark on the line plot. Therefore, 5 students had $\frac{3}{4}$ of a cup of leftover lemonade.

Question: How many students had $1\frac{1}{2}$ of a cups of leftover lemonade?
There are 3 x's above the $1\frac{1}{2}$ mark on the line plot. Therefore, 3 students had $1\frac{1}{2}$ of a cups of leftover lemonade.

Question: Is there any conclusion that could be made from the results shown in the line plot? Yes, one conclusion is that about $\frac{2}{3}$ of the students drank half or more than half of their glass of lemonade.

Chapter 8   Measurements and Line Plots

Read each problem carefully and answer the questions that follow. (DOK 2)

Each student in science class was asked to measure out 10 grams of salt and dissolve it in $\frac{1}{8}$ liter of water. Each student was told to measure the amount of time it took to dissolve all of the salt. The results are shown in the line plot below.

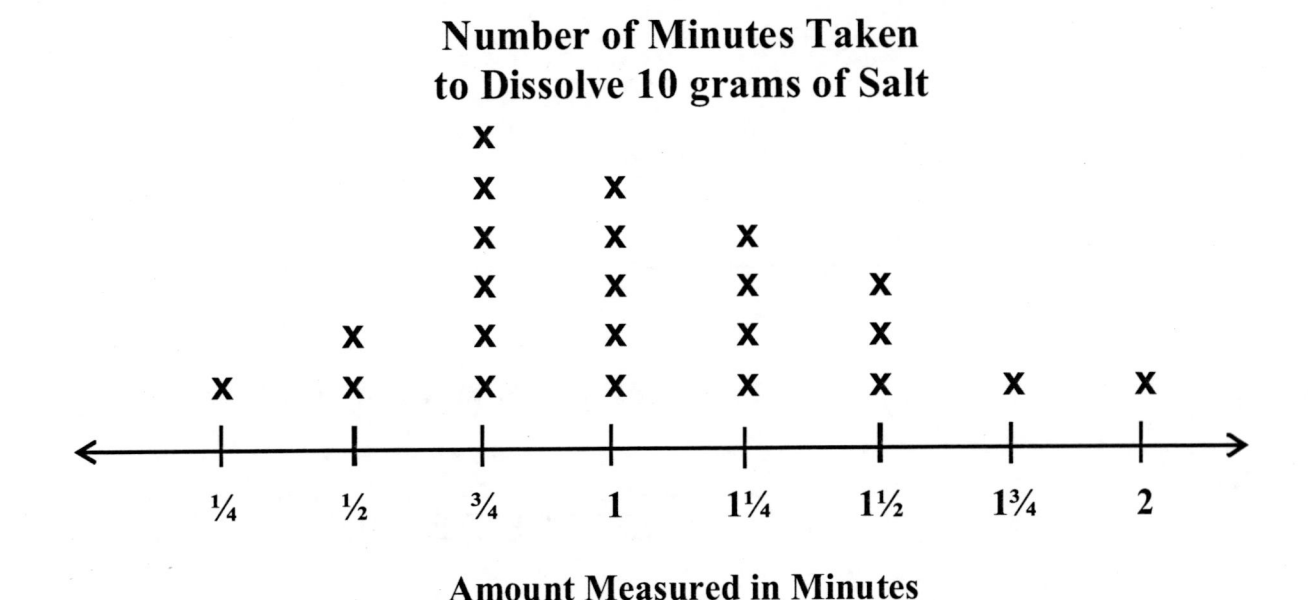

Number of Minutes Taken to Dissolve 10 grams of Salt

```
                    x
                    x    x
                    x    x    x
                    x    x    x    x
               x    x    x    x    x
          x    x    x    x    x    x    x    x
     x    x    x    x    x    x    x    x
   ──┼────┼────┼────┼────┼────┼────┼────┼──→
    ¼    ½    ¾    1   1¼   1½   1¾   2
```

Amount Measured in Minutes

Key
Each x represents 1 student.

1. How many students had results of $\frac{1}{2}$ minute?

2. How many students had results of $1\frac{1}{2}$ minute?

3. How many students had results of $\frac{3}{4}$ minute?

4. What is the total number of students included in the line plot?

8.10 Line Plots (DOK 2)

The students in Mr. Sullivan's class were asked to measure the width of the classroom using three different methods.

Method 1: Measure the length of your foot, then walk heal to toe across the classroom, and then multiply the length of your foot by the number of "feet" across the classroom.

Method 2: Measure using a one foot ruler, and kneeling on the floor, measure the width of the classroom moving the ruler along the floor.

Method 3: Measure using a metal tape measure. Ask one student to hold the tab on the tape measure against the wall on one side of the classroom, while another student walk to the other side of the classroom and read the tape's measure.

The results of the measures are shown in the line plot below.

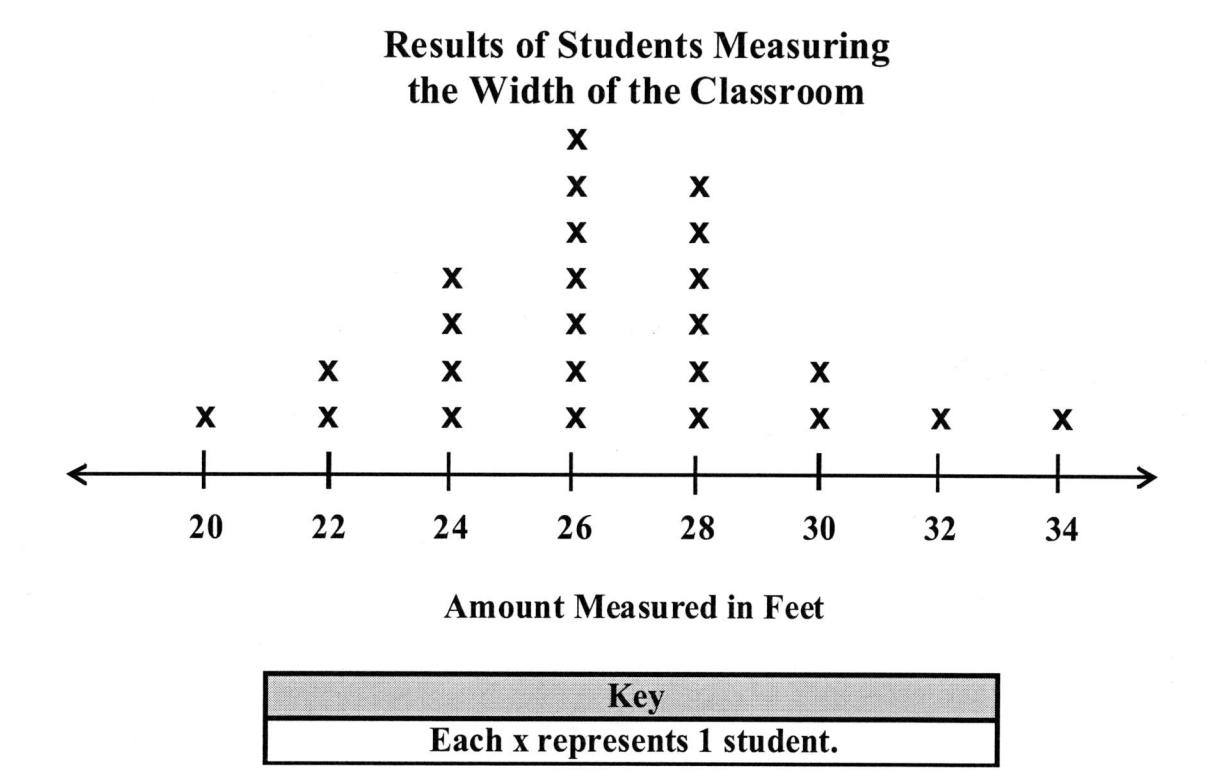

Results of Students Measuring the Width of the Classroom

```
                        x
                        x   x
                        x   x
                 x      x   x
                 x      x   x
            x    x      x   x    x
       x    x    x      x   x    x    x    x
       |    |    |      |   |    |    |    |
       20   22   24     26  28   30   32   34
```

Amount Measured in Feet

Key
Each x represents 1 student.

5. What measure yielded the most results?

6. How many students found the width of the classroom measured 30, 32, or 34 feet?

7. How many students found the width of the classroom measured 24 feet?

8. True or false: Based on the results in the line plot, the width of the classroom likely measures 26–28 feet across?

Chapter 8 Measurements and Line Plots

Another experiment in measuring was tried by Mr. Silverman's class. The students were asked to measure the distance from the front of the school to the town's library down the street.

Method 1: Strap a pedometer to your ankle and program the length of one step into the pedometer. Then walk from the school to the library. The pedometer will measure mileage by multiplying the number of steps by the length of one step to give you your mileage.

Method 2: Ask an adult in your family to drive from the front of the school to the library. Write down the number of miles on the car's odometer at the start of the drive and when you reach the library. Subtract the two numbers to come up with the measure.

Method 3: Working in pairs, use a 100 foot length of rope and, starting at the front of the school, one student will hold still while another student will walk 100 feet. Keeping track of the number of times you walk 100 feet, alternate, so the other student will walk 100 feet, and so on, until you reach the library. Count up the number of 100 foot lengths. There are 5,280 feet in a mile. Using division, find the measure.

Results of Measuring the Distance from the Front of the School to the Library

```
                              x
                              x
                              x
                        x     x
                        x     x     x
                        x     x     x
                        x     x     x
                        x     x     x     x
                  x     x     x     x     x     x
            x     x     x     x     x     x     x
  ←――――┼―――――┼―――――┼―――――┼―――――┼―――――┼―――――┼―――――┼――――→
        ¼     ½     ¾     1    1¼    1½    1¾     2
```

Amount Measured in Miles

Key
Each x represents 1 student.

9. How many students found the distance to measure $1\frac{1}{4}$ miles?

10. How many students found the distance to measure $1\frac{1}{2}$ miles?

11. How many students found the distance to measure $1\frac{3}{4}$ or 2 miles?

12. True or false: Based on the results in the line plot, the distance between the front of the school and the library is likely $1 - 1\frac{1}{4}$ miles?

8.11 Going Deeper into Measurements and Line Plots (DOK 3)

Solve the multi-step problems below. Show your work. (DOK 3)

1. Andrew pours 600 mL of water into an empty 2 liter bottle. He then adds 1 liter of whole milk into the same 2 liter bottle and mixes the milk and water together. He then pours 240 mL of the milk/water mixture into each of 5 glasses. How much of the milk/water mixture is left in the 2 liter bottle?

2. Michah's grandmother canned 16 gallons of peaches into quart jars. She gave away 8 quarts of the peaches to relatives. She gave away 17 quarts of the peaches to friends and neighbors. She used 2 quarts of the peaches to make peach cobbler. How many gallons of peaches does Michah's grandmother have left?

3. Walter wants to make 2 gallons of punch for a family reunion. He starts with 2 quarts of lemonade, $\frac{3}{4}$ of a gallon of apple juice, 2 pints of lime sherbert, and fills up the rest of the 2 gallon punch bowl with lemon-lime soda. How much lemon-lime soda did Walter use?

4. Emma made a table of how long her beaded strings are. Beads are sized by the width of the bead measured in millimeters.

Kind of Beaded String	Length
Red 5 mm beads	65 cm
Light Blue 8 mm beads	120 cm
Dark Blue 10 mm beads	40 cm
Pink 3 mm beads	110 cm
Purple 6 mm beads	80 cm

How many <u>meters</u> of beads does Emma have in all? Express your answer in decimal form.

5. Mr. Taylor enjoys a morning jog on most days. The table below shows the number of meters he jogged on Monday through Friday this week.

Day	Number of Meters
Monday	2,200
Tuesday	3,500
Wednesday	1,750
Thursday	2,400
Friday	3,000

How many <u>kilometers</u> did Mr. Taylor jog in all for the five days? Express your answer in decimal form.

6. Mrs. Verdon has a recipe for beet soup that calls for 18 ounces of beets. She gives the recipe to the school cafeteria and they decide to make the beet soup for the Wednesday school lunches. Her recipe makes 6 servings. The school will be making 1,200 servings. How many pounds of beets will the school need to buy? Express your answer in pounds and ounces.

7. A salt shaker in the Pearson house has 40 grams of salt in it. The Pearson family uses the salt as follows: Mr. Pearson uses 25 grams to make a bar-b-que sauce. Mrs. Pearson uses 3 grams to make cookies. Annie Pearson adds 21 grams of salt to the shaker. Mack Pearson uses 4 grams on his salsa. The Pearson cat knocks over the salt shaker and 2 grams spill out. Grandma Pearson uses 22 grams of salt in a fermented vegetable recipe. Annie Pearson adds another 35 grams of salt to the shaker. How much salt is in the shaker now?

Chapter 8   Measurements and Line Plots

Chapter 8 Review

Fill in the blanks below with the appropriate English or metric conversions. (DOK 2)

1. Two gallons equals _____ cups.

2. 4.2 L equals _____ mL.

3. 3 yards equals _____ inches.

4. 6,800 m equals _____ kilometers.

5. 36 oz equals _____ pounds.

6. 73,000 mg equals _____ kg.

Solve the following problems. (DOK 2)

7. 120 m = _____ km

8. 9 g = _____ mg

9. 0.02 kL = _____ L

10. 15 mg = _____ g

11. 15 cm = _____ mm

12. 5 L = _____ mL

13. 0.005 kg = _____ g

14. 55 mL = _____ L

15. 30 cm = _____ m

16. Mandy's new baby sister weighs $8\frac{1}{2}$ pounds. When Mandy was born, she weighed 116 ounces. What is the difference in their two birth weights? Give your answer in ounces.

17. William measured the length of his partly used pencil at 11 cm. When the pencil was new, it measured 0.16 meters. What is the difference in the measure between the new pencil and the partly used pencil? Give your answer in centimeters.

18. Jalisa took a new gallon of milk and poured three 8 ounce glasses for herself and her two friends. How many cups of milk are left in the gallon?

Chapter 8 Review

Mrs. Watson told her fifth grade class to measure the volume, to the rim, of identical glass jars. The students were asked to use one of the 3 methods below, giving their answers in ounces. The results of the class experiment are shown in the line plot below. **(DOK 2)**

Method 1: Use a measuring cup and pour $\frac{1}{4}$ cup of water into the jar until it is full. Count how many $\frac{1}{4}$ cup measures you pour into the jar and multiply by 2 ounces.

Method 2: Fill the jar to the rim, and then pour the water out of the jar into three measuring cups. Write down the measures, add them up, and then convert the result to ounces to get the answer.

Method 3: Time yourself while drinking $\frac{1}{2}$ cup of water. Then time yourself drinking all of the water in the jar. Divide the total time by the number of seconds per $\frac{1}{2}$ cup of water to come up with your answer.

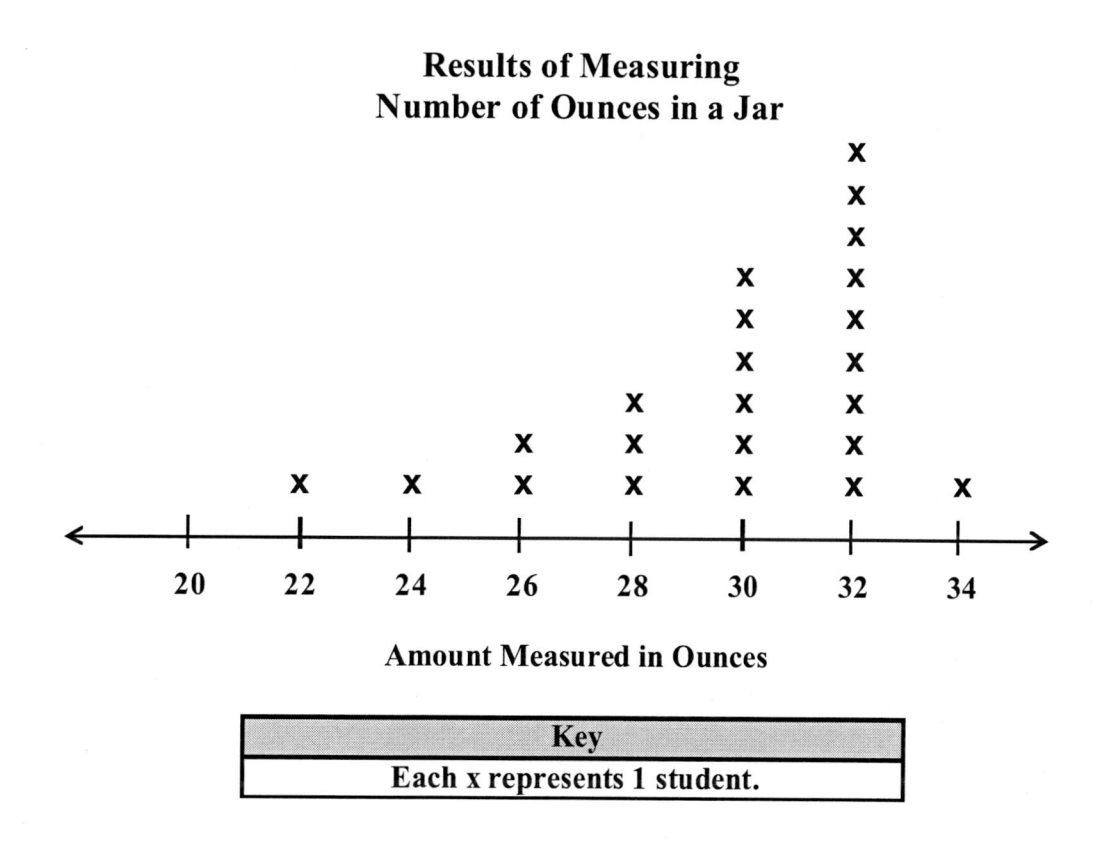

Results of Measuring Number of Ounces in a Jar

```
                                              x
                                              x
                                              x
                                    x         x
                                    x         x
                                    x         x
                          x         x         x
                    x     x         x         x
     x     x     x  x     x         x     x
   ──┼─────┼─────┼──┼─────┼─────┼─────┼─────┼──→
     20    22    24 26    28    30    32    34
```

Amount Measured in Ounces

Key
Each x represents 1 student.

19. Based on the results shown in the line plot, what is the likely number of ounces in each jar?

20. How many students found the jar held 28 ounces?

21. How many students found the jar held 20 − 26 ounces?

Chapter 8 Measurements and Line Plots

Chapter 8 Test

1 How many yards are there in one mile?

A 1,760
B 5,280
C 1,670
D 5,820

(DOK 2)

2 Which metric measurement would you use to measure the mass of a vitamin?

A millimeter
B milligram
C kilometer
D vitimeter

(DOK 1)

3 Estimating the amount of ice tea needed for a party, Clare's mom figured 16 ounces per person. There are 18 people coming to the party. How many gallon(s) of ice tea will Clare's mom make?

A $3\frac{3}{4}$ gallons
B 4 gallons
C $4\frac{1}{4}$ gallons
D $2\frac{1}{4}$ gallons

(DOK 2)

4 What is the equivalent of 1 kilometer?

A 1,000 meters
B 100 meters
C 100 centimeters
D 100,000 hectometers

(DOK 1)

5 Mrs. Daniels bought 18 yards of fabric to make new curtains. How many feet is 18 yards?

A 6 feet
B 36 feet
C 54 feet
D 12 feet

(DOK 2)

6 How many cups are there in $\frac{1}{2}$ gallon of apple juice?

A 16 cups
B 12 cups
C 8 cups
D 20 cups

(DOK 2)

7 Enrique is measuring drops of sweetener for his lemonade. What unit of measurement are used with drops of sweetener?

A liters
B milliliters
C cups
D pints

(DOK 1)

8 Alma's grandmother canned 8 quarts of tomatoes. What is the equivalent in pints?

A 2 pint
B 4 pints
C 12 pints
D 16 pints

(DOK 2)

9 Cindy just ran 3,000 meters. What is the equivalent in kilometers?

A 3 kilometers
B $\frac{1}{2}$ kilometer
C 30 kilometers
D 300 kilometers

(DOK 1)

10 Kenny is 2 yards and 3 inches tall. How tall is Kenny in feet?

A $4\frac{1}{2}$
B $5\frac{1}{4}$
C $6\frac{1}{4}$
D 7

(DOK 2)

11 Jessica is making muffins and needs 4 ounces of buttermilk for the recipe. How many cups of buttermilk should Jessica use?

A 2

B 1

C $\frac{12}{4}$

D $\frac{1}{2}$

(DOK 2)

12 Meggan is 1.3 meters tall. Her dog, Fishy, is 42 centimeters tall. What is the difference in height between Meggan and her dog?

A 43.3 cm

B 88 cm

C 1,342 cm

D 172 cm

(DOK 2)

13 If you were measuring the thickness of a coin, what is the unit of measurement you would likely use?

A kilometers

B meters

C centimeters

D millimeters

(DOK 1)

14 254 milligrams is equal to:

A 254 grams

B 25.4 grams

C 2.54 grams

D 0.254 grams

(DOK 1)

15 How many pints are in $3\frac{1}{4}$ gallons?

A 13

B 15

C 24

D 26

(DOK 2)

16 How many feet are in $\frac{1}{2}$ mile?

A 5,280

B 1,000

C 2,640

D 500

(DOK 2)

17 Which unit of the metric system is used to measure the mass of a person?

A kilograms

B milligrams

C centigrams

D grams

(DOK 1)

18 Frederick is making tacos for 12 people. If he uses 2 ounces of meat for each taco, how many pounds of meat will he use?

A $1\frac{1}{2}$ lb

B 2 lb

C $2\frac{1}{4}$ lb

D 3 lb

(DOK 2)

19 Rick used 2 lengths of rope to make a rope swing. Each piece measured 12 feet. How many yards of rope did Rick use in all?

A 12

B 9

C 8

D 6

(DOK 2)

20 Carl was measuring the length of a grasshopper he found in the park. Which unit of measurement should Carl use?

A kilometers

B centimeters

C millimeters

D meters

(DOK 1)

Chapter 9
Plane Geometry

This chapter covers the following CC 5 standards:

| Number and Operations - Fractions | 5.NF.4 |
| Geometry | 5.G.3, 5.G.4 |

9.1 Polygons (DOK 1)

Polygon: A closed plane figure with 3 or more straight sides. Polygons can be grouped by the total number of sides and angles they have.

Types of Polygons

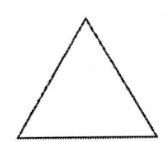

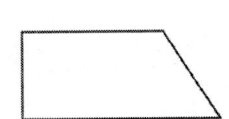

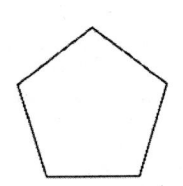

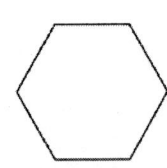

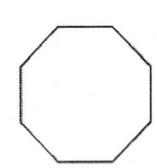

Triangle	Quadrilateral	Pentagon	Hexagon	Octagon
3 sides, 3 angles	4 sides, 4 angles	5 sides, 5 angles	6 sides, 6 angles	8 sides, 8 angles

9.2 Quadrilaterals and Their Properties (DOK 1)

A **quadrilateral** is a polygon with 4 sides and 4 angles.

Shape	Type of quadrilateral	Lines	Sides	Angles
	General quadrilateral	4 straight lines	4	4
	Square	2 pairs of parallel lines that are opposite each other	All equal sides	All right angles
	Rectangle	2 pairs of parallel lines that are opposite each other	2 pairs of equal sides	All right angles
	Parallelogram	2 pairs of parallel lines that are opposite each other	2 pairs of equal sides	4
	Rhombus	2 pairs of parallel lines that are opposite each other	4 equal sides	4
	Trapezoid	1 pair of parallel lines that are opposite each other	4	4

Chapter 9 Plane Geometry

A **parallelogram** is a quadrilateral in which both pairs of opposite lines are parallel. Based upon the chart, we can determine that the following figures are parallelograms.

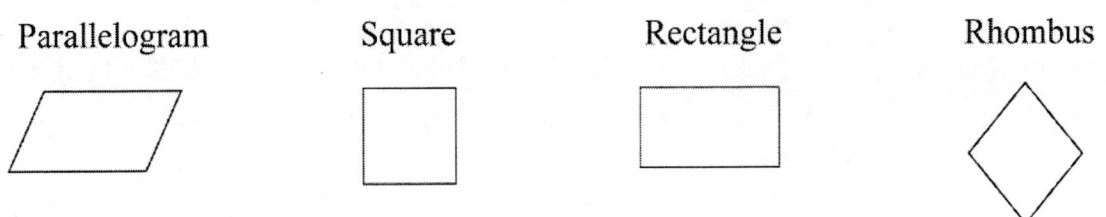

Parallelogram   Square   Rectangle   Rhombus

9.3 Identifying Figures (DOK 1, 2)

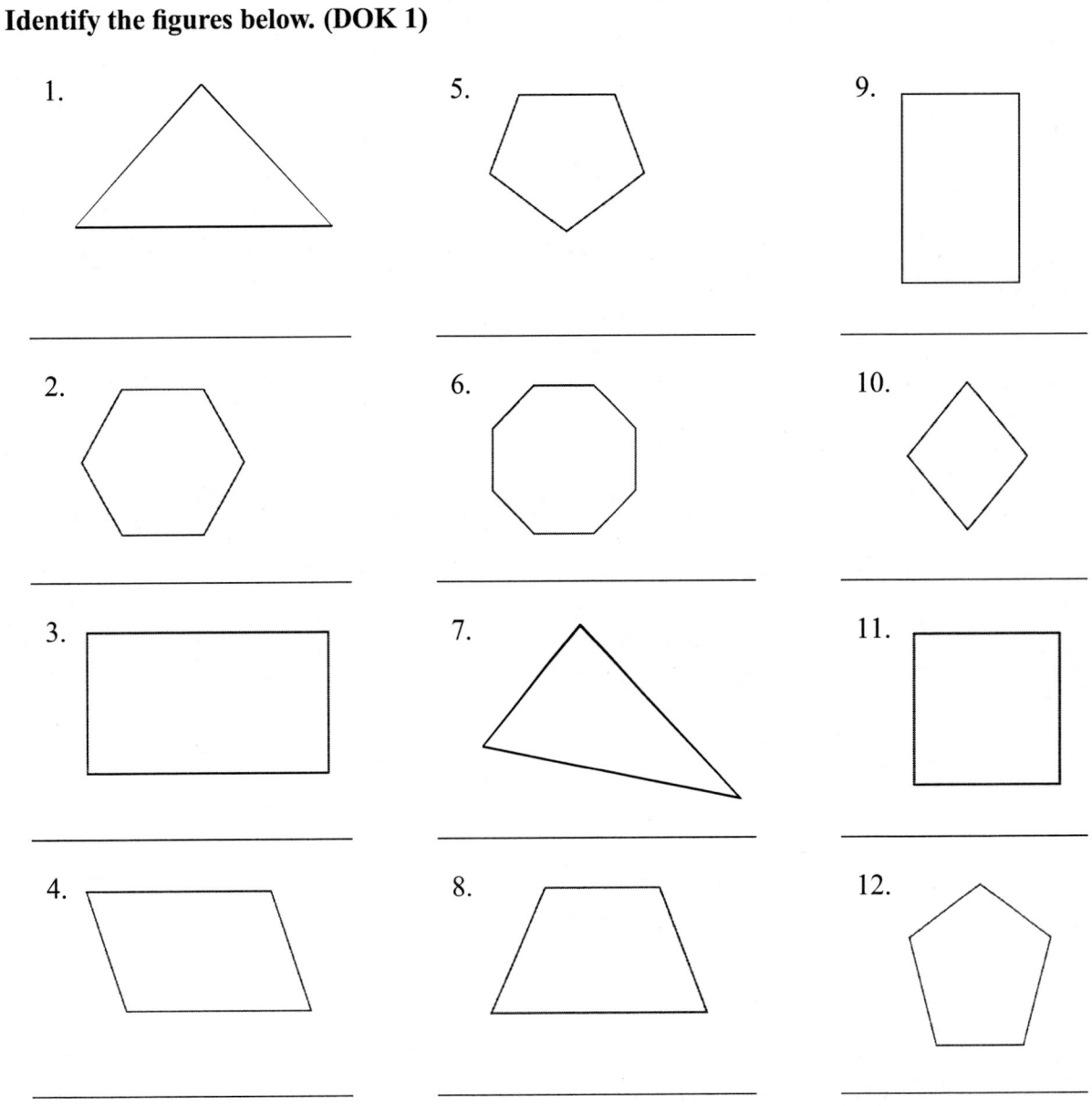

Identify the figures below. (DOK 1)

1.
2.
3.
4.
5.
6.
7.
8.
9.
10.
11.
12.

9.3 Identifying Figures (DOK 1, 2)

Identify the polygons in the descriptions below. (DOK 2)

13. Which of the following has four right angles?

    trapezoid   pentagon   rectangle   triangle

14. Which polygon has six sides?

    rhombus   hexagon   pentagon   trapezoid

15. Which quadrilateral has four equal sides and angles?

    trapezoid   square   rectangle   triangle

16. Which polygon most resembles a stop sign?

    trapezoid   square   octagon   pentagon

17. Which polygon most resembles the Zippy Oil Change sign below?

    trapezoid   square   octagon   pentagon

18. Which polygon has two pairs of parallel sides?

    trapezoid   parallelogram   triangle   pentagon

19. Which polygon has only one pair of parallel sides?

    trapezoid   square   octagon   pentagon

Chapter 9 Plane Geometry

9.4 Properties of Shapes (DOK 2)

Shapes are defined by their sides and angles. In describing the sides of a shape, we include the number of sides, whether the sides are the same size or not, and whether the sides are parallel or perpendicular to each other. In describing the angles of a shape, we include the number of angles, if they are the same size, and if the angles are right angles or other kinds of angles.

Parallel lines are two or more lines that are equal distance to each other. An easy example of this is the equal sign "=." The lines of an equal sign are equal distance apart and never intersect.

Perpendicular lines are two lines that meet or cross each other at a right angle (90°). A plus sign is an example of perpendicular lines "+."

In describing each of the twelve shapes below, only "regular" shapes will be referenced. For instance, a pentagon has five sides. In this section, we will only be referring to a regular pentagon with five equal sides, not a pentagon with different lengths of sides. The same goes for the rest of the shapes.

Triangle - 3 sides and 3 angles. The 3 sides and angles may be equal as in the case of an equilateral triangle, 2 sides and angles may be equal as in the case of an isosceles triangle, or no sides and angles equal as in the case of a scalene triangle.

Square - All 4 sides are equal, 2 pairs of sides parallel to each other. All 4 angles are equal and are right angles. The diagonal lines connecting the opposite corners of a square meet at right angles.

Rectangle - 2 pairs of equal and parallel sides, all 4 angles are right angles. One pair of sides may be longer than the other pair of sides. The diagonal lines connecting the opposite corners of a rectangle meet making one pair of equal acute angles and one pair of equal obtuse angles.

Trapezoid - 2 sides are made up of 1 pair of parallel lines that are opposite each other, and 2 sides that are not parallel to each other.

Rhombus - All 4 sides are equal. 2 pairs of sides that are parallel to each other. 2 pairs of equal angles. The diagonal lines connecting the opposite corners of a rhombus meet at right angles.

Parallelogram - 2 pairs of parallel lines that are opposite each other, making up two pairs of equal sides and 2 pairs of equal angles. The diagonal lines connecting the opposite corners of a parallelogram meet making one pair of equal acute angles and one pair of equal obtuse angles.

Pentagon - 5 sides. A regular pentagon has 5 equal sides and 5 equal angles.

Hexagon - 6 sides. A regular hexagon has 6 equal sides and 6 equal angles, 3 pairs of sides are parallel and opposite each other.

Heptagon - 7 sides. A regular heptagon has 7 equal sides and 7 equal angles.

Octagon - 8 sides. A regular octagon has 8 equal sides and 8 equal angles, 4 pairs of sides are parallel and opposite each other.

Nonagon - 9 sides. A regular nonagon has 9 equal sides and 9 equal angles.

Decagon - 10 sides. A regular decagon has 10 equal sides and 10 equal angles, 5 pairs of sides are parallel and opposite each other.

9.4   Properties of Shapes (DOK 2)

Choose the shape that best fits the following descriptions. (DOK 2)

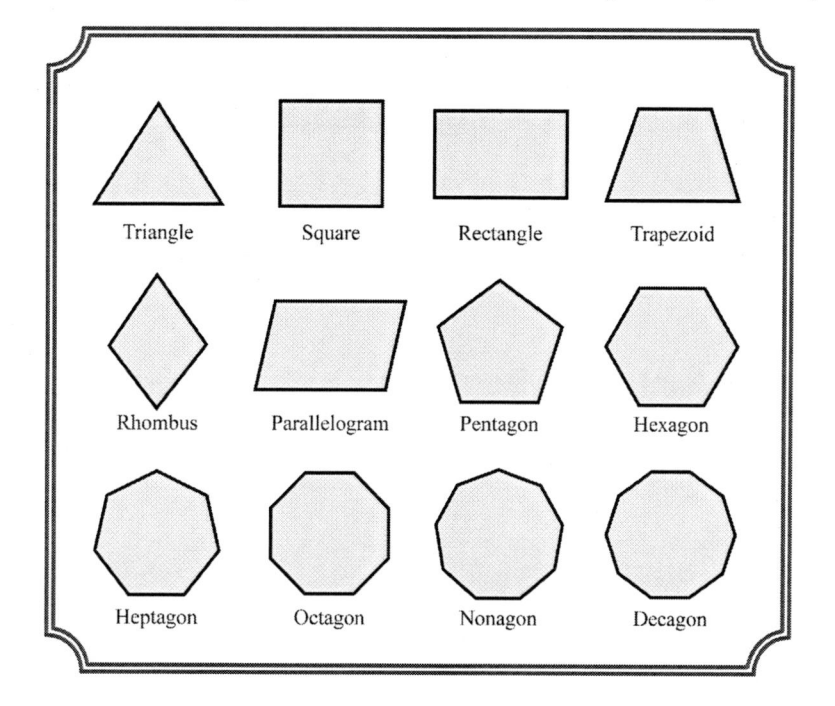

Angle Review	
Acute	Less than 90°
Obtuse	More than 90°
Right	Exactly 90°

1. A shape that has 7 sides and angles.

2. A shape that has 1 set of parallel sides, 4 sides, and 4 angles.

3. A shape with 3 sides and 3 angles.

4. A shape with 9 sides and 9 angles.

5. A shape with 4 equal sides and 2 pairs of parallel sides, which are perpendicular to each other.

6. A shape with 3 pairs of parallel sides and 6 angles.

7. Two shapes with 2 pairs of parallel sides, 2 pairs of equal sides, and 2 pairs of equal angles.

8. A shape with 4 pairs of parallel sides and 8 equal angles.

9. A shape with 4 equal sides and 4 right angles.

10. A shape with 5 pairs of parallel sides and 10 angles.

11. A shape with 2 pairs of parallel sides, one pair may be longer than the other, and four 90° angles.

12. A shape with 5 equal sides and 5 equal angles.

Chapter 9   Plane Geometry

Read each question and use the shape(s) shown to answer the questions. (DOK 2)

13. Which of the following is a plane figure with 5 sides and 5 obtuse angles?

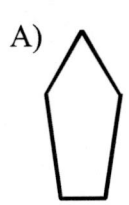

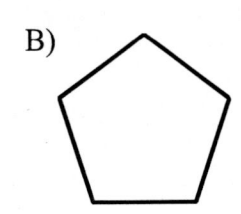

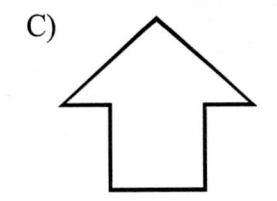

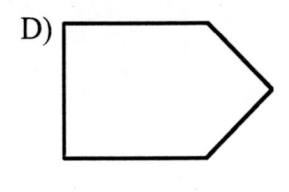

    A)   B)   C)   D)

14. In trapezoid *KLMNO*, which side is parallel to side $\overline{LM}$?

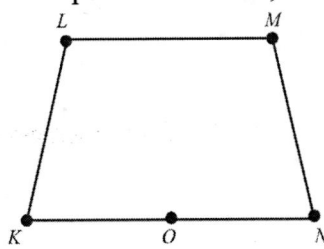

15. Which two line segments on the pentagon below are perpendicular?

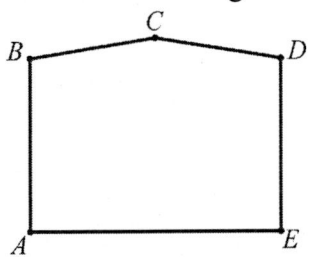

    A) $\overline{BA}$ and $\overline{DE}$ B) $\overline{BA}$ and $\overline{AE}$ C) $\overline{BA}$ and $\overline{BC}$ D) $\overline{BC}$ and $\overline{DE}$

16. Which of these polygons is always a regular polygon?

    A) rectangle   B) triangle   C) square   D) octagon

17. Marcy drew a diagonal inside a quadrilateral and made two right triangles. What kind of quadrilateral did Marcy use to draw the triangles?

    A) parallelogram   B) rhombus   C) trapezoid   D) rectangle

18. Which of the following polygons has only one pair of parallel sides?

    A) trapezoid   B) square   C) rectangle   D) pentagon

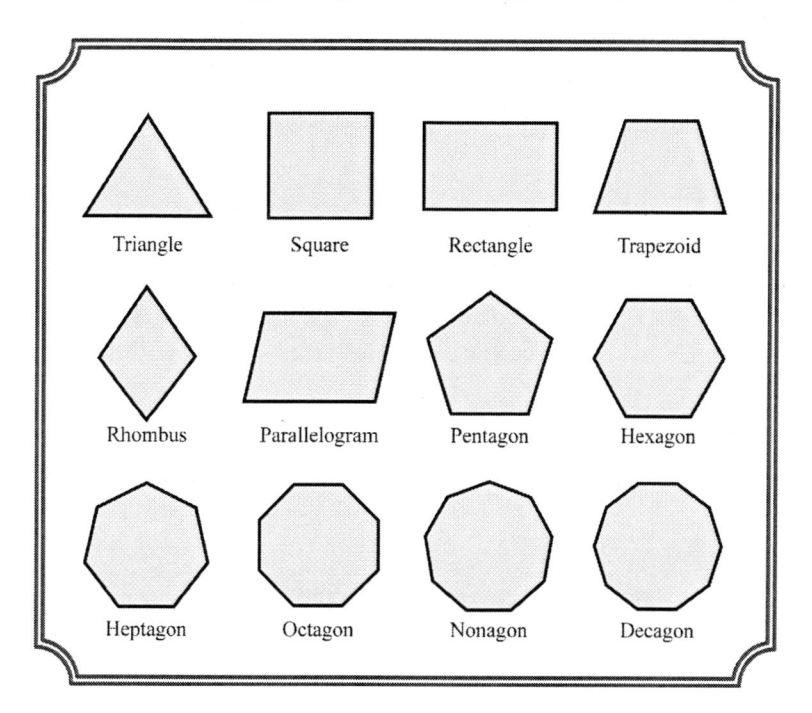

9.5 Putting Shapes in Categories (DOK 2)

Put shapes into the categories provided, using the chart of shapes above. (DOK 2)

1. List all the shapes with 4 sides.

2. List all the shapes with at least 2 acute angles.

3. List all the shapes with at least 2 right angles.

4. List all the shapes with at least 1 obtuse angle.

5. List all the shapes with four right angles.

6. List all the shapes that have at least one set of parallel lines.

Answer the true/false questions. If the answer is false, explain why. (DOK 2)

7. <u>True or false:</u> All rectangles have four right angles. Squares have four right angles. Therefore, squares are rectangles.

8. <u>True or false:</u> All squares have four right angles and four equal sides. Rectangles have four right angles. Therefore, rectangles are squares.

9. <u>True or false:</u> Trapezoids have four sides. Therefore, trapezoids are squares.

10. <u>True or false:</u> Rectangles have two sets of parallel sides. Therefore, rectangles are parallelograms.

11. <u>True or false:</u> Rhombuses have 2 acute angles and 2 obtuse angles. Therefore, rhombuses are trapezoids.

Chapter 9   Plane Geometry

9.6   Area of Rectangles (DOK 2)

The area of a rectangle is equal to the number of square units that can cover the rectangle. The units may be any measure: in, ft, yd, mile, mm, cm, m, km, etc. If the measure is not shown on the graphic, use the word "units.".

Example 1:

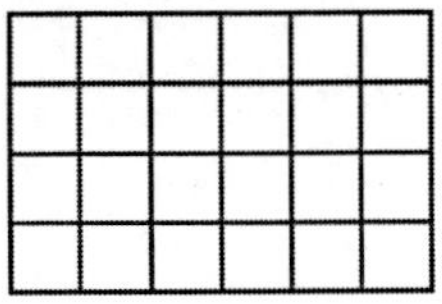

$A = lw$ (length × width)
$A = 4 \times 6$
$A = 24$ square units
If a rectangle has an area of 24 square units, it means that it will take 24 squares that are 1 unit on each side to cover the area that is 4 units × 6 units on each side.

Example 2:

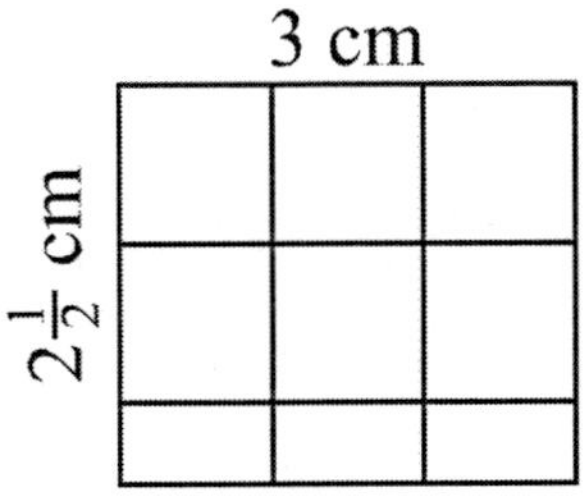

$A = lw$
$A = 2\frac{1}{2} \times 3$
$A = 7\frac{1}{2}$ square cm

Refer to the "About AR" on page xi!

Find the area of the rectangles. Express your answer in the square units given. (DOK 2)

1. $4\frac{1}{2}$ in × $2\frac{1}{2}$ in

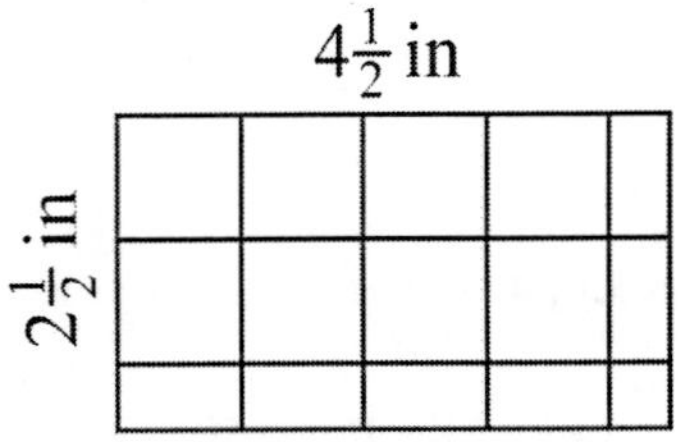

2. 3 meters × $3\frac{1}{4}$ meters

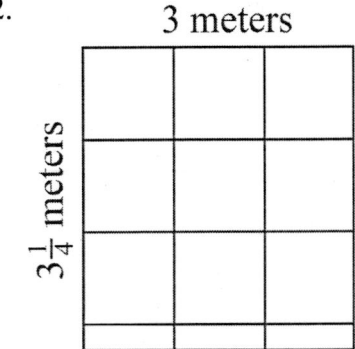

3. $4\frac{3}{4}$ feet × $3\frac{1}{4}$ feet

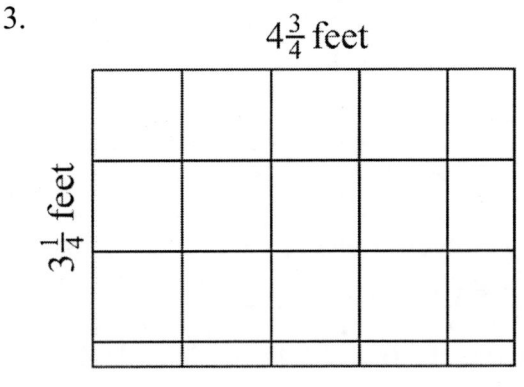

4. $7\frac{1}{2}$ units × 6 units

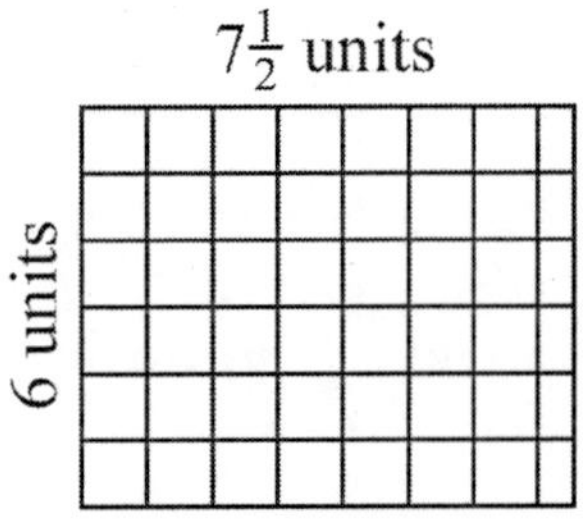

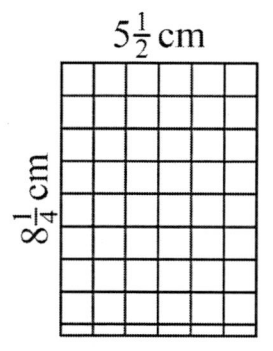

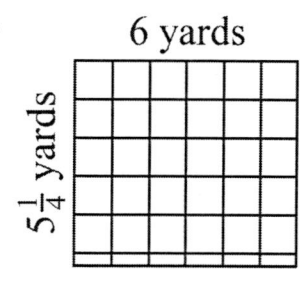

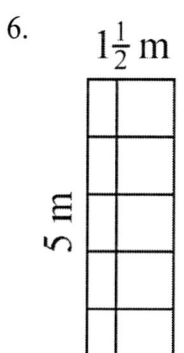

9.6 Area of Rectangles (DOK 2)

5. $5\frac{1}{2}$ cm, $8\frac{1}{4}$ cm

6. $1\frac{1}{2}$ m, 5 m

7. 6 yards, $5\frac{1}{4}$ yards

8. $4\frac{1}{2}$ in, $5\frac{1}{4}$ in

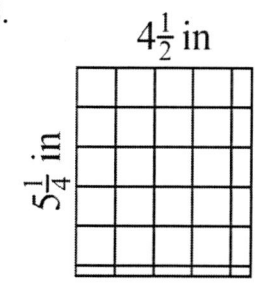

9. $14\frac{1}{2}$ feet, $10\frac{1}{2}$ feet

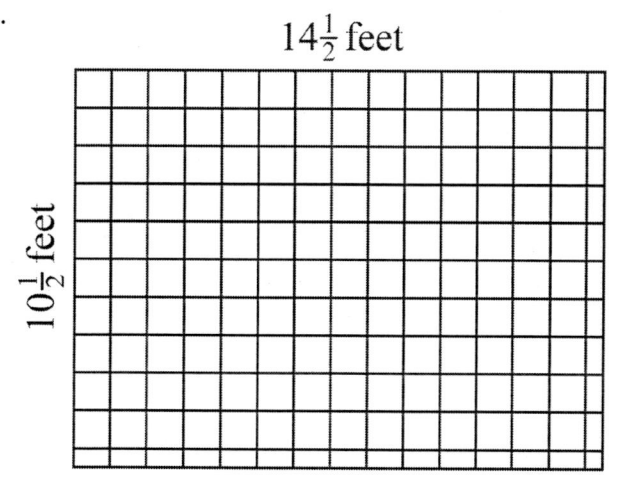

10. 25 inches, $6\frac{1}{4}$ inches

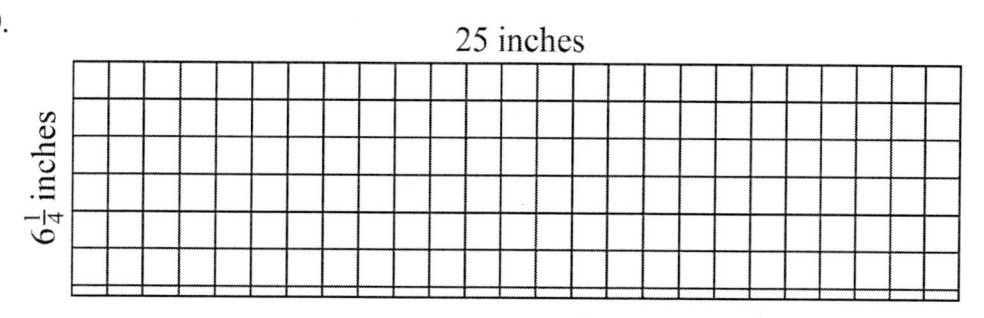

Chapter 9   Plane Geometry

9.7   Going Deeper into Plane Geometry (DOK 3)

Construct drawings according to the instructions below. (DOK 3)

1. Draw a figure that has the following properties:

   A) Draw a grid modeling a rectangle that has an area of 36 square inches.
   B) Attached to the bottom of the rectangle from Part A, draw a pentagon.
   C) Attached to the right side of the grid in part A, draw a triangle.
   D) Attached to the bottom of the triangle in part C, draw a hexagon.

2. Draw a figure that has the following properties:

   A) Draw any quadrilateral.
   B) Attached to the right of the quadrilateral in part A, draw a triangle
   C) Attached to the right side of the triangle in part B, draw a hexagon.
   D) Attached to the bottom of the hexagon in part C, draw a square.

3. Draw a figure that has the following properties:

   A) Draw a grid modeling a rectangle that has an area of 20 square cm.
   B) Attached to the bottom of the rectangle from Part A, draw an octagon.
   C) Attached to the bottom of the octagon in part B, draw a square.
   D) Attached to the left of the square in part C, draw a pentagon.

4. Draw a figure that has the following properties:

   A) Draw a rhombus.
   B) Attached to the lower, right side of the rhombus in part A, draw a square.
   C) Attached to the square in part B, attach a triangle.
   D) Attached to the bottom of the triangle in part C, draw a trapezoid.

5. Draw a figure that has the following properties:

   A) Draw a grid modeling a rectangle that has an area of 12 square inches.
   B) Attached to the bottom of the rectangle from part A, draw a parallelogram.
   C) Attached to the right side of the parallelogram in part B, draw a triangle.
   D) Attached to the bottom of the triangle in part C, draw a square.

Chapter 9 Review

Name the shape. (DOK 1)

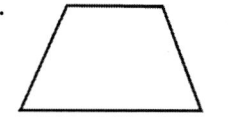

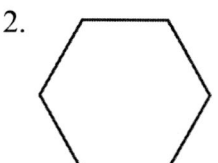

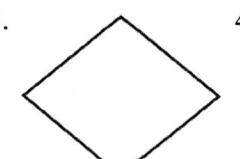

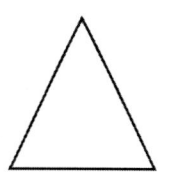

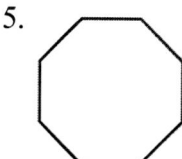

1. (trapezoid) 2. (hexagon) 3. (rhombus) 4. (triangle) 5. (octagon)

Answer the questions about polygons. (DOK 2)

6. Which two of the following have perpendicular lines?

    A) square   B) regular pentagon   C) rectangle   D) regular octagon

7. Which two of the following have at least two sets of parallel lines?

    A) pentagon   B) rectangle   C) trapezoid   D) regular octagon

8. True or false: All four sided figures can be categorized as rectangles.

9. Which two of the following have exactly four sides?

    A) square   B) pentagon   C) trapezoid   D) octagon

10. Which of the following has at least 2 acute angles?

    A) square   B) pentagon   C) trapezoid   D) rhombus

11. What is the area of a 7 centimeter square?

12. What is the area of the figure below? Round your answer to the nearest hundredth.

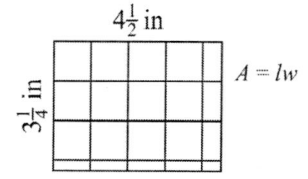

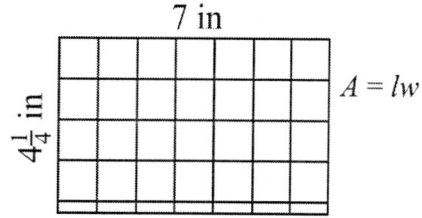

    $4\frac{1}{2}$ in, $3\frac{1}{4}$ in, $A = lw$

13. What is the area of the figure below?

    7 in, $4\frac{1}{4}$ in, $A = lw$

Chapter 9 Plane Geometry

Chapter 9 Test

1 Find the area of the figure below.

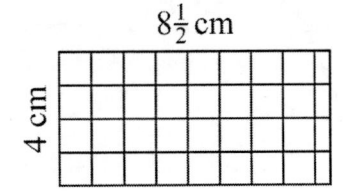

$8\frac{1}{2}$ cm
4 cm

- **A** 32 square cm
- **B** 34 square cm
- **C** 36 square cm
- **D** $34\frac{1}{2}$ square cm

(DOK 2)

2 Which quadrilateral appears to have all perpendicular lines?

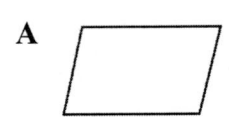

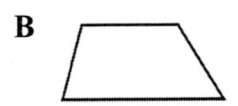

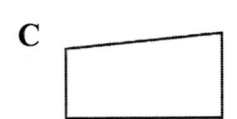

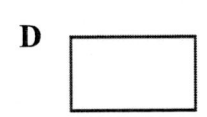

A

B

C

D

(DOK 2)

3 Which shape has 6 sides and 6 angles?

- **A** trapezoid
- **B** hexagon
- **C** pentagon
- **D** octagon

(DOK 1)

4 What is the area of the figure below?

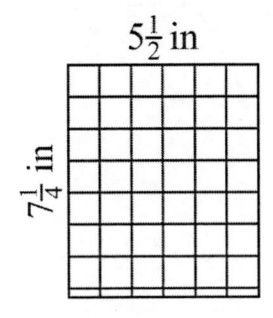

$5\frac{1}{2}$ in
$7\frac{1}{4}$ in

- **A** $38\frac{3}{4}$ square in
- **B** $38\frac{7}{8}$ square in
- **C** $39\frac{7}{8}$ square in
- **D** $35\frac{7}{8}$ square in

(DOK 2)

5 Which 2 shapes have a total of 13 angles?

- **A** rectangle and pentagon
- **B** heptagon and square
- **C** hexagon and pentagon
- **D** pentagon and octagon

(DOK 2)

6 What is the area of a 9 inch square?

- **A** 81 square inches
- **B** 18 square inches
- **C** 9 square inches
- **D** 36 square inches

(DOK 2)

Chapter 9 Test

7 Mrs. Taylor tilled up the soil in her backyard to prepare for a vegetable garden. The garden measures 12 feet by 14 feet. What is the area of the vegetable garden?

A 168 square feet

B 56 square feet

C 158 square feet

D 66 square feet

(DOK 2)

8 Which of the figures below is a trapezoid?

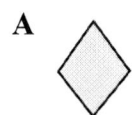

A

B

C

D

(DOK 1)

9 Which shape below has at least 1 set of parallel sides?

A triangle

B rectangle

C heptagon

D pentagon

(DOK 2)

10 Which group of figures below are all quadrilaterals?

A square, rectangle, pentagon, trapezoid

B trapezoid, square, rhombus, hexagon

C octagon, rectangle, trapezoid, square

D rectangle, trapezoid, rhombus, square

(DOK 2)

11 Which sentence about shapes is true?

A All rectangles are squares.

B All squares are rectangles.

C All quadrilaterals are rectangles.

D All trapezoids are squares.

(DOK 2)

12 Which pair of figures both have four right angles?

A square and pentagon

B rectangle and trapezoid

C rhombus and parallelogram

D square and rectangle

(DOK 2)

13 Which pair of figures is made up of only obtuse angles?

A octagon and hexagon

B parallelogram and rhombus

C trapezoid and octagon

D rectangle and square

(DOK 2)

14 Which shape has 8 sides and 8 angles?

A hexagon

B rhombus

C octagon

D square

(DOK 2)

15 The drawing below has a triangle, a pentagon, and what other shape?

A rectangle

B square

C rhombus

D parallelogram

(DOK 3)

Copyright © American Book Company

Chapter 10
Solid Geometry

This chapter covers the following CC 5 standards:

Measurement and Data	5.MD.3, 5.MD.4, 5.MD.5

10.1 Understanding Volume (DOK 2)

Measurement of volume is expressed in cubic units such as cubic inches, cubic feet, cubic meters, cubic centimeters, or cubic millimeters. The volume of a solid is the number of cubic units that can fit in the solid.

Example 1: How many 1 cubic units will it take to fill up the rectangular solid below?

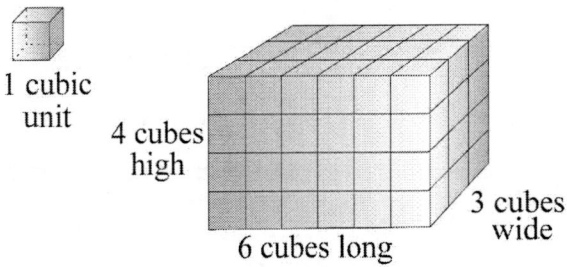

1 cubic unit
4 cubes high
6 cubes long
3 cubes wide

To find the volume, you can count the number of cubes, or a shorter method is to multiply the length times the width times the height.

Volume of a rectangular solid = length × width × height $(V = lwh)$

$$V = 6 \times 3 \times 4 = 72 \text{ cubic units}$$

Find the volume of the blocked figures below. Use the formula $V = lwh$. (DOK 2)

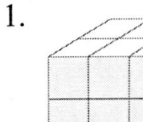

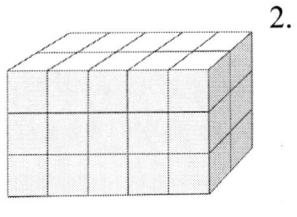

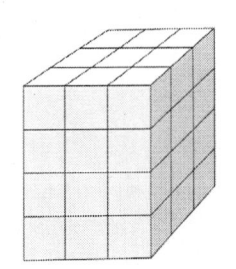

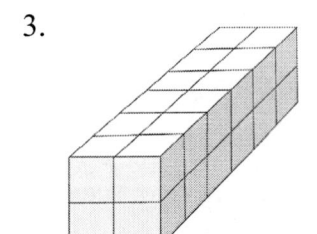

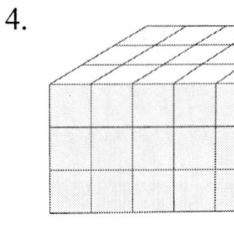

1.
2.
3.
4.

10.2 Volume of Rectangular Prisms (DOK 2)

You can calculate the volume (V) of a rectangular prism (box) by multiplying the measure of the length (l) by the width (w) by the height (h), as expressed in the formula $V = l \times w \times h$. The volume can also be expressed as $V = b \times h$. The b is the base, which is length × width. First find the base and then multiply by the height.

Example 2: Find the volume of the box pictured here:

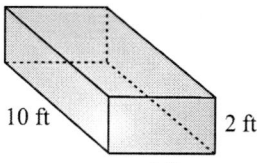

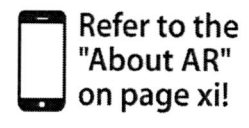

Step 1: Insert measurements from the figure into the formula. $V = 10 \times 4 \times 2$

Step 2: Multiply to solve. $10 \times 4 \times 2 = 80$ cubic feet

Find the volume of the following rectangular prisms (boxes). (DOK 2)

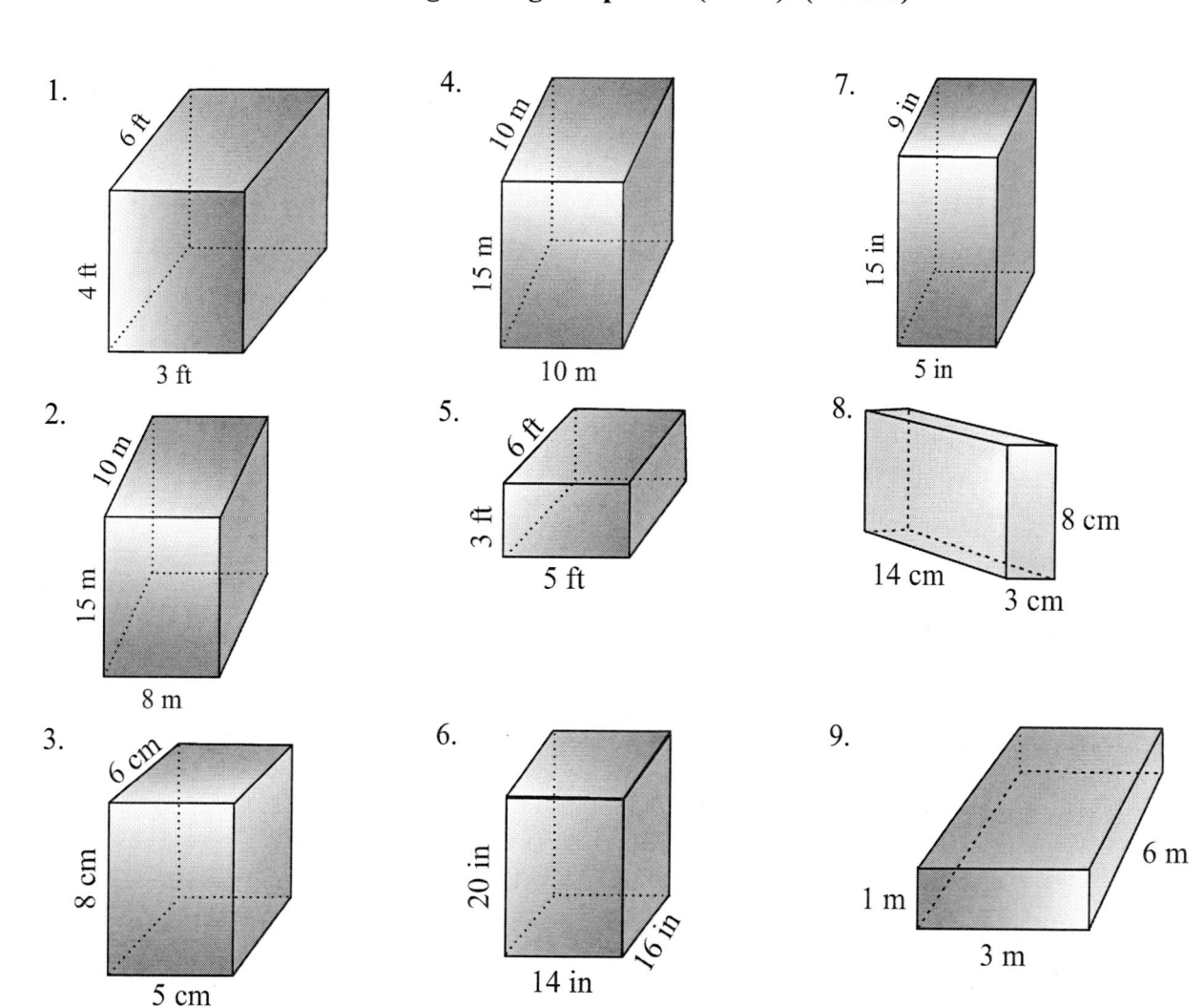

1. 6 ft, 4 ft, 3 ft
2. 10 m, 15 m, 8 m
3. 6 cm, 8 cm, 5 cm
4. 10 m, 15 m, 10 m
5. 6 ft, 3 ft, 5 ft
6. 20 in, 14 in, 16 in
7. 9 in, 15 in, 5 in
8. 8 cm, 14 cm, 3 cm
9. 6 m, 1 m, 3 m

Copyright © American Book Company

133

Chapter 10   Solid Geometry

10.3   Volume of Cubes (DOK 2)

A **cube** is a special kind of rectangular prism (box). Each side of a cube has the same measure. So, the formula for the volume of a cube is $V = s \times s \times s$.

Example 3:   Find the volume of the cube at right:

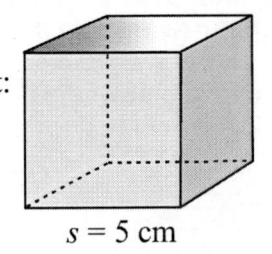

$s = 5$ cm

Step 1:   Insert measurements from the figure into the formula. $V = 5 \times 5 \times 5$
Step 2:   Multiply to solve. $5 \times 5 \times 5 = 125$ cubic centimeters

Answer each of the following questions about cubes. (DOK 2)

1. If a cube is 3 centimeters on each edge, what is the volume of the cube?

2. If the measure of the edge is doubled to 6 centimeters on each edge, what is the volume of the cube?

3. What if the edge of a 3 centimeter cube is tripled to become 9 centimeters on each edge? What will the volume be?

4. How many cubes with edges measuring 3 centimeters would you need to stack together to make a solid 12 centimeter cube?

5. What is the volume of a 2-centimeter cube?

6. Jerry built a 2-inch cube to hold his marble collection. He wants to build a cube with a volume 8 times larger. How much will each edge measure?

Find the volume of the following cubes. (DOK 2)

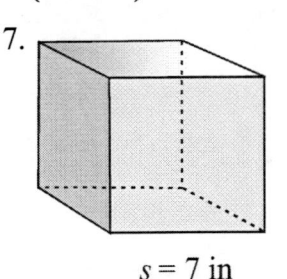

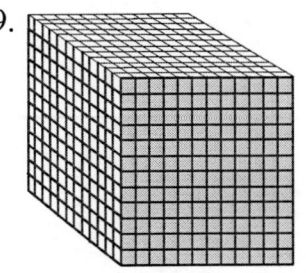

7. $s = 7$ in

8. 4 ft, 4 ft, 4 ft

9. $s = 1$ foot
How many cubic inches are in a cubic foot? A foot is 12 inches long.

134   Copyright © American Book Company

10.4 Volume of Compound Figures (DOK 3)

Finding the volume of compound figures is the same as finding the volume of rectangular prisms, except there are two or more parts to the figure. Find the volume of each part and add all the parts together.

Example 4: Find the volume of the compound figure below.

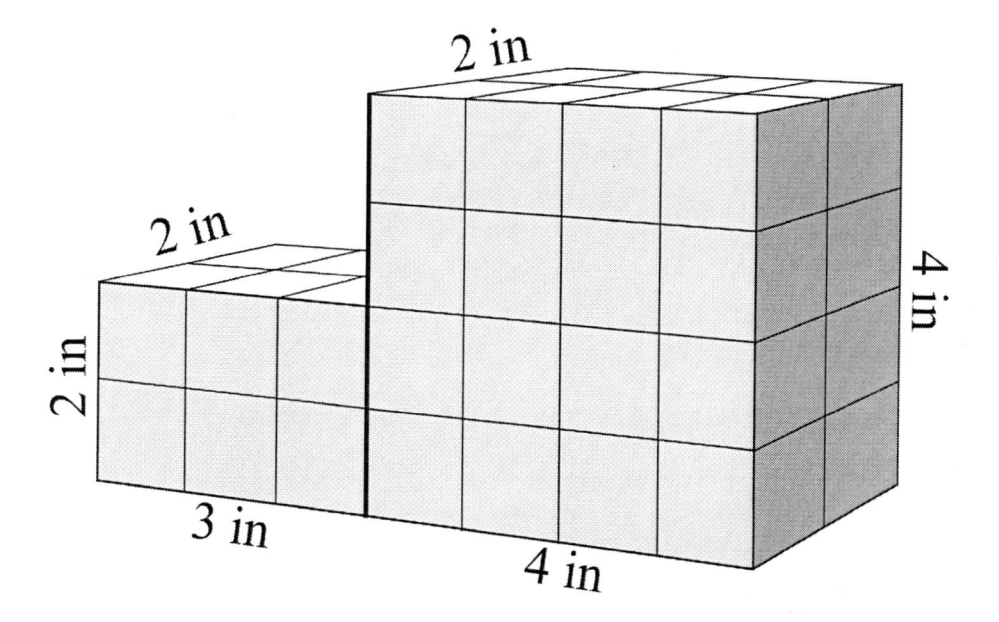

Step 1: Find the volume of the left side of the compound figure.
$2 \times 2 \times 3 = 12$ cubic inches

Step 2: Find the volume of the right side of the compound figure.
$2 \times 4 \times 4 = 32$ cubic inches

Step 3: Add the volumes of the two sides of the compound figure.
$12 + 32 = 44$ cubic inches

Answer: 44 cubic inches

Copyright © American Book Company

Chapter 10   Solid Geometry

Find the volume of the compound figures below. (DOK 3)

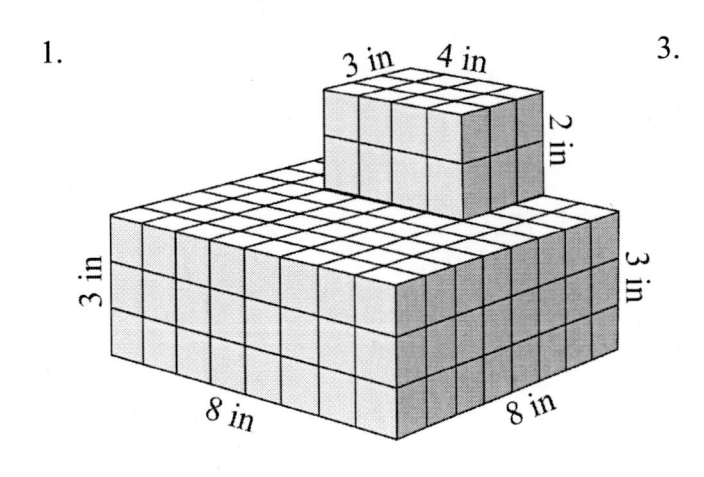

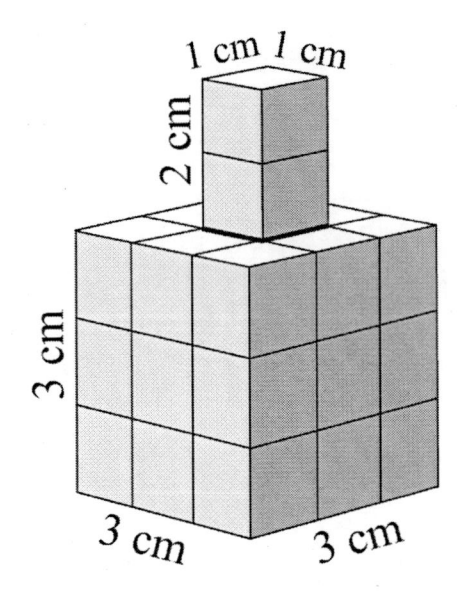

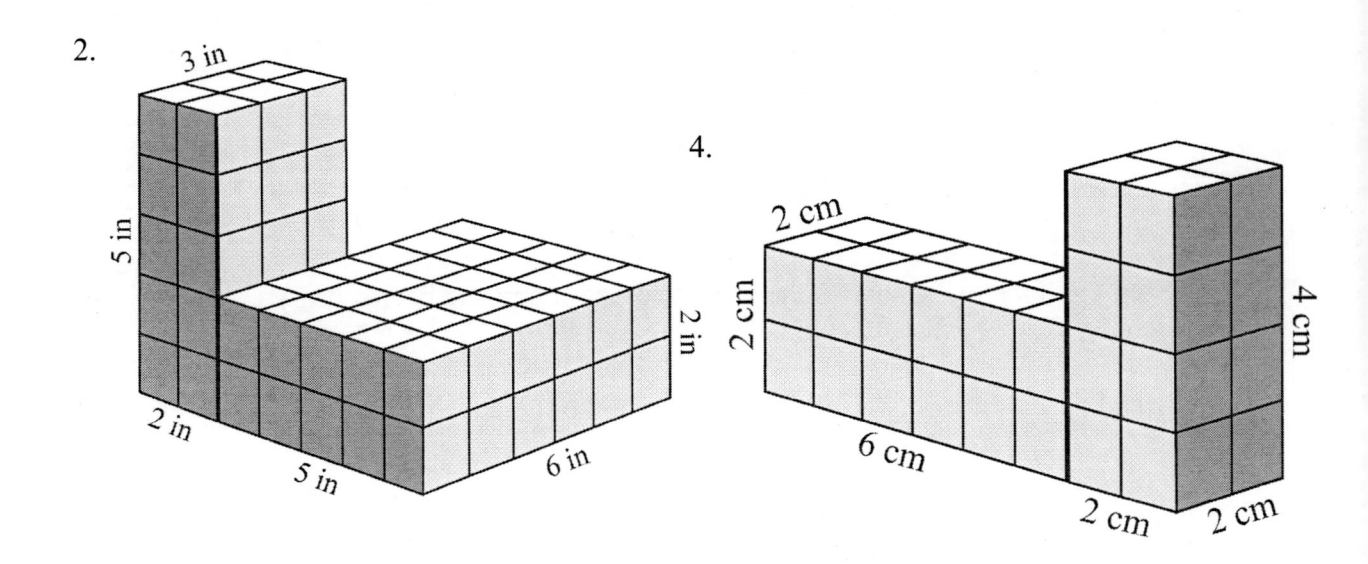

1.

2.

3.

4.

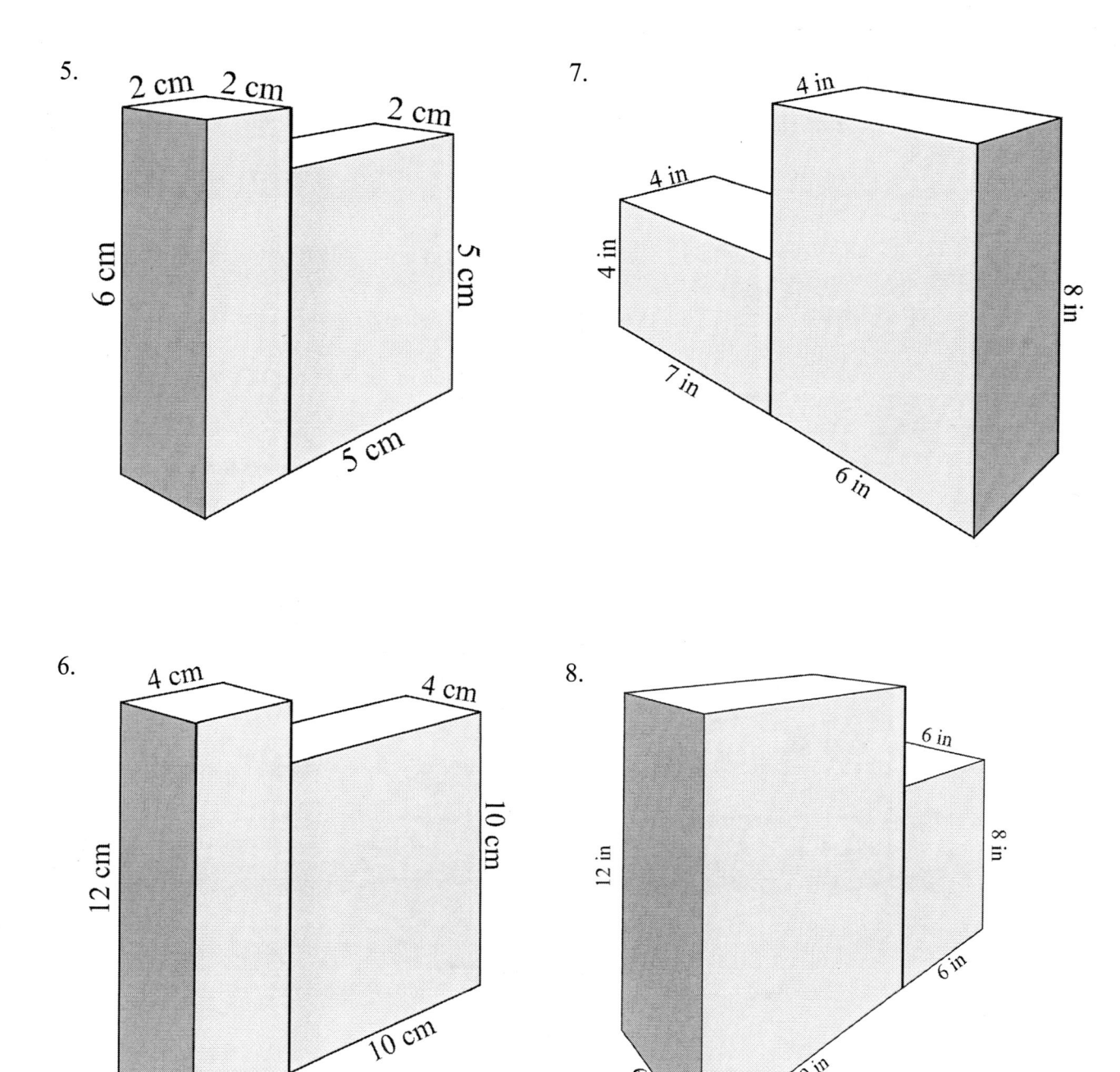

10.4 Volume of Compound Figures (DOK 3)

5. 2 cm, 2 cm, 2 cm, 6 cm, 5 cm, 5 cm

7. 4 in, 4 in, 4 in, 7 in, 8 in, 6 in

6. 4 cm, 4 cm, 12 cm, 10 cm, 10 cm, 4 cm

8. 6 in, 12 in, 8 in, 6 in, 6 in, 9 in

Chapter 10   Solid Geometry

10.5   Real–World Volume Problems (DOK 2, 3)

Find the volume of these real world examples. Answer the questions with each graphic. (DOK 2, 3)

1. What is the volume in units cubed of the building to the left?

2. What is the volume in units cubed of the building below?

3. What is the difference in the number of units cubed between the two buildings?

4. What is the total number of units cubed for both buildings?

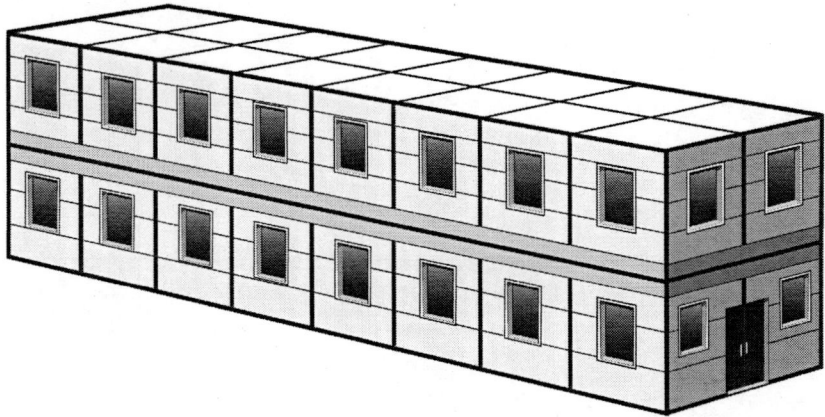

5. What is the volume in inches cubed of the box of Rice Crunchies?

6. What is the volume in inches cubed of the box of Corn Flakies?

7. What is the difference in volume between the two boxes of cereal?

8. What is the total volume of the two boxes of cereal?

Chapter 10 Review

Find the volume of the compound figures below. (DOK 3)

1. (3 ft, 10 ft, 2 ft, 9 ft, 6 ft, 3 ft)

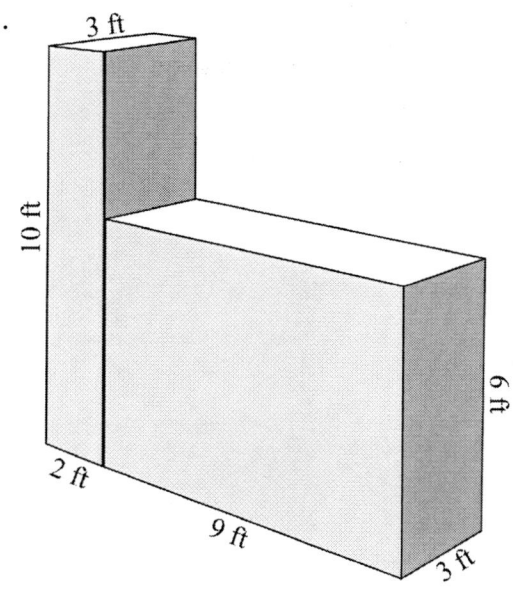

2. (4 cm, 5 cm, 4 cm, 8 cm, 4 cm, 5 cm)

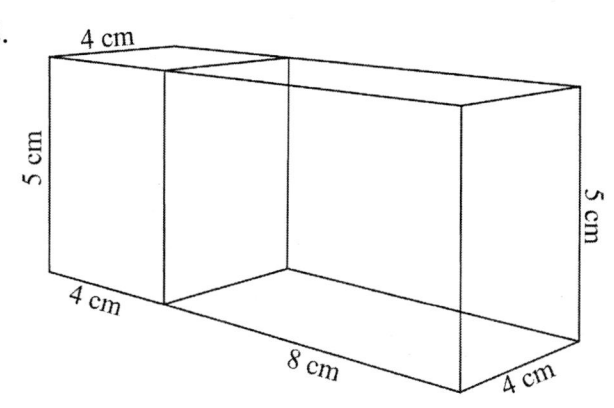

Find the volume of the rectangular prisms below. $V = lwh$ **or** $V = bh$ **(DOK 2)**

3.

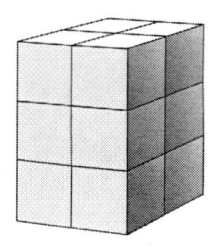

4. (8 cm × 3 cm × 6 cm)

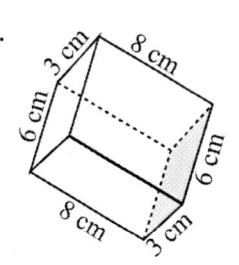

5. (4 in × 1 in × 5 in)

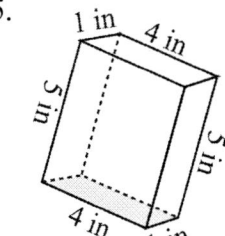

Find the volume of the cubes below. $V = s \times s \times s$ **(DOK 2)**

6. (3 in)

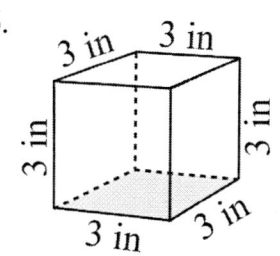

7. (5 in)

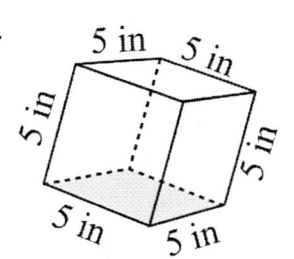

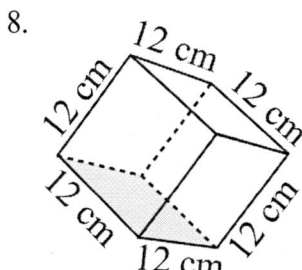

8. (12 cm)

Chapter 10   Solid Geometry

Find the volume of the real world figures. Then answer the questions that follow. (DOK 2, 3)

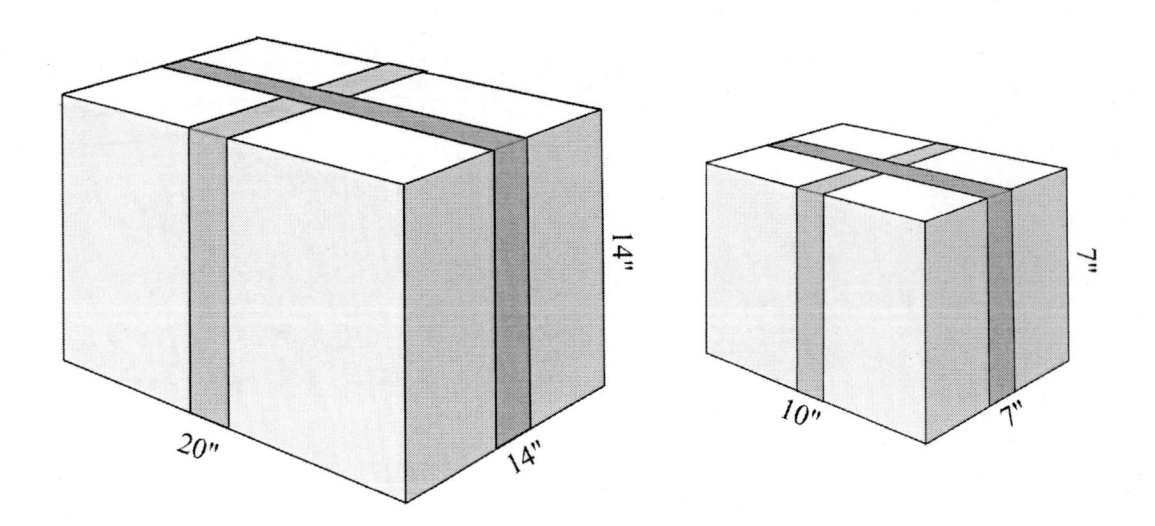

9. What are the volumes of the two boxes separately?

10. What is the difference in the volume of the two boxes?

11. If you filled the volume of the smaller box with sand and emptied it into the larger box over and over until the larger box was full, how many times would you fill the smaller box with sand?

Find the volume of the boxes below. Then answer the questions that follow. (DOK 2)

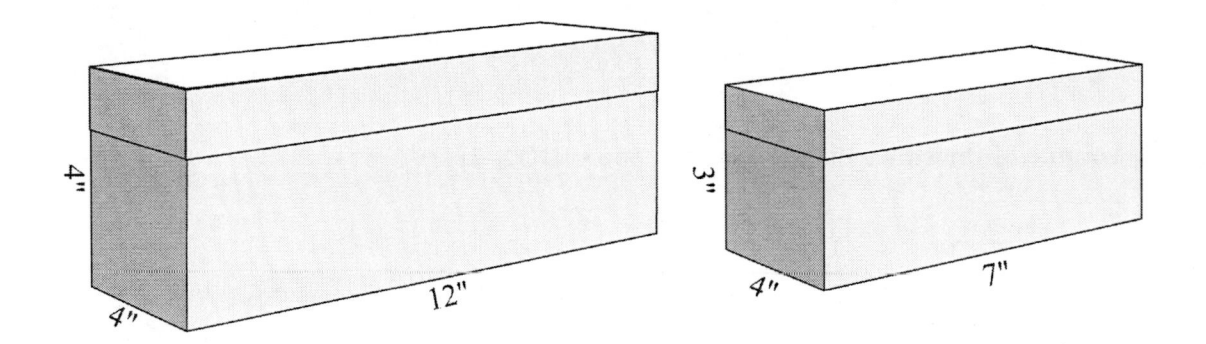

12. What are the volumes of the two boxes separately?

13. What is the difference in the volume of the two boxes?

Chapter 10 Test

1 Find the volume of the figure below.

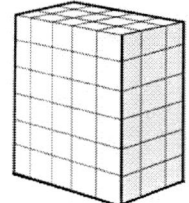

$V = lwh$

- **A** 15 cubic units
- **B** 18 cubic units
- **C** 88 cubic units
- **D** 90 cubic units

(DOK 2)

2 Find the volume of the cube below.

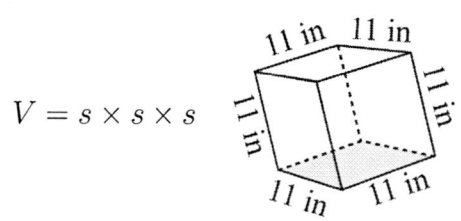

$V = s \times s \times s$

- **A** 121 cubic inches
- **B** 1,331 cubic inches
- **C** 111 cubic inches
- **D** 1,221 cubic inches

(DOK 2)

3 Find the volume of the cube below.

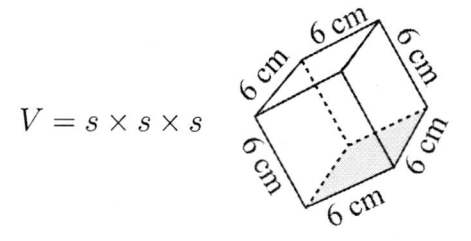

$V = s \times s \times s$

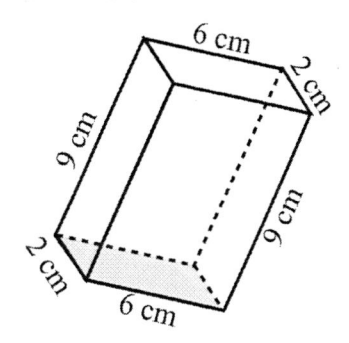

- **A** 36 cubic centimeters
- **B** 66 cubic centimeters
- **C** 246 cubic centimeters
- **D** 216 cubic centimeters

(DOK 2)

4 Find the volume of the rectangular prism below.

$V = lwh$

- **A** 108 cubic centimeters
- **B** 98 cubic centimeters
- **C** 76 cubic centimeters
- **D** 116 cubic centimeters

(DOK 2)

5 Find the volume of the rectangular prism below.

$V = lwh$

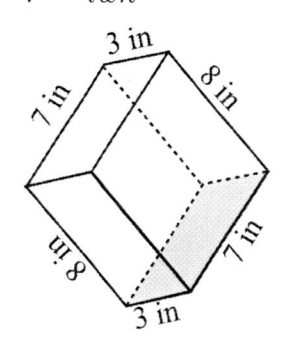

- **A** 116 cubic inches
- **B** 148 cubic inches
- **C** 168 cubic inches
- **D** 174 cubic inches

(DOK 2)

Chapter 10   Solid Geometry

6 Find the volume of the compound figure below.

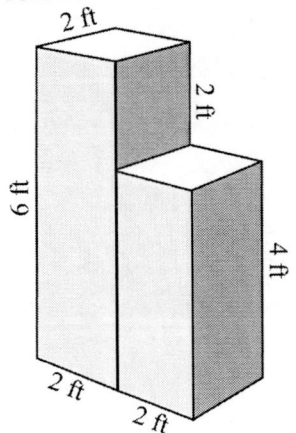

A 16 cubic feet

B 24 cubic feet

C 40 cubic feet

D 42 cubic feet

(DOK 2)

7 What is the volume of the compound figure below?

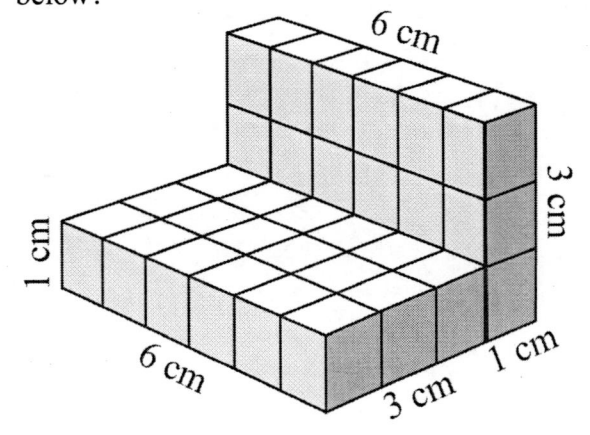

A 32 cubic feet

B 36 cubic feet

C 42 cubic feet

D 46 cubic feet

(DOK 3)

8 What is the sum of the volume of the two show boxes below?

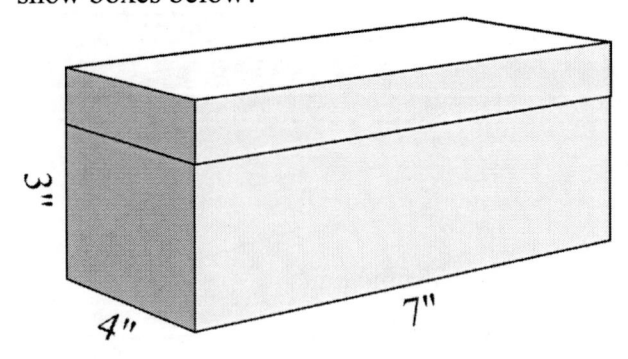

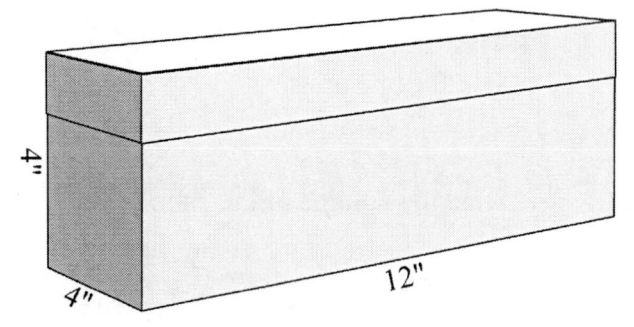

A 276 cubic inches

B 192 cubic inches

C 284 cubic inches

D 246 cubic inches

(DOK 2)

9 What is the difference in volume of the two boxes in problem #8, plus the volume of a third box that measures 2" by 6" by 9"?

A 116 cubic inches

B 108 cubic inches

C 192 cubic inches

D 216 cubic inches

(DOK 3)

Chapter 11
Introduction to Graphing

This chapter covers the following CC 5 standards:

| Geometry | 5.G.1, 5.G.2 |

11.1 The Coordinate Grid and Ordered Pairs (DOK 1)

A **number line** allows you to graph points with one value. For instance, the number 2 is graphed on the number line below.

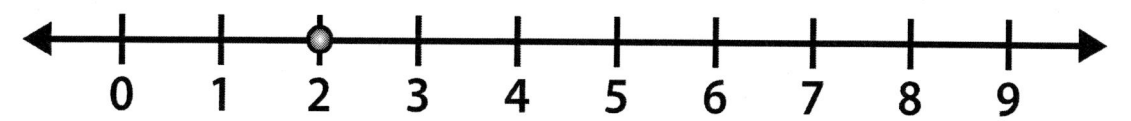

A **coordinate grid** allows you to graph points with two values. A coordinate grid has two number lines, the horizontal line called the x-axis and the vertical line called the y-axis. The point where the x and y axes intersect is called the **origin**. Each point on the coordinate grid is designated by an **ordered pair** of coordinates. The ordered pairs are enclosed in parentheses, with the x-axis named first and the y-axis named second: (x, y). These values are called **coordinates**. An easy way to remember which number comes first is the axes are in alphabetical order. X comes before Y in the alphabet, and horizontal comes before vertical in the alphabet.

To identify a coordinate pair, simply count from the origin, $(0, 0)$.

Example 1: The ordered pair in the coordinate grid below is $(2, 3)$.

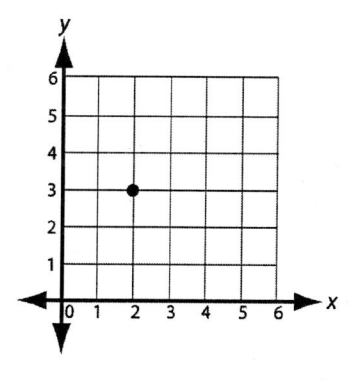

The point is above the number 2 on the x-axis, and the point is across from the 3 on the y-axis. Therefore, the coordinate pair is $(2, 3)$.

Copyright © American Book Company

Chapter 11 Introduction to Graphing

Example 2: The ordered pair in the coordinate grid below is (1, 4).

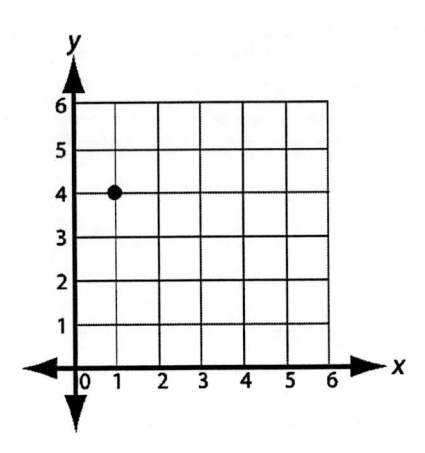

The point is above the number 1 on the x-axis and the point is across from the number 4 on the y-axis. Therefore, the coordinate pair is (1, 4).

If the ordered pair has a zero in it, the marker will land on either the horizontal or vertical number line. If the ordered pair begins with a zero, such as (0, 5), the marker for the ordered pair will land on the vertical number line. If the ordered pair has a zero in the second place, such as (8, 0), the marker for the ordered pair will land on the horizontal number line.

Locate the point on the coordinate grids below and name the ordered pair. (DOK 1)

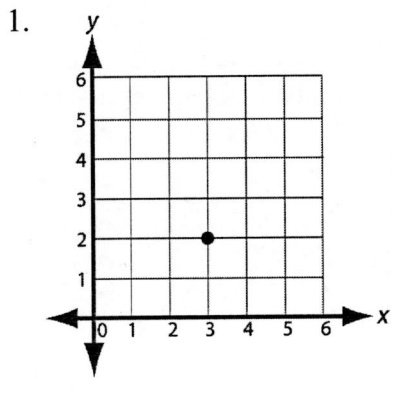

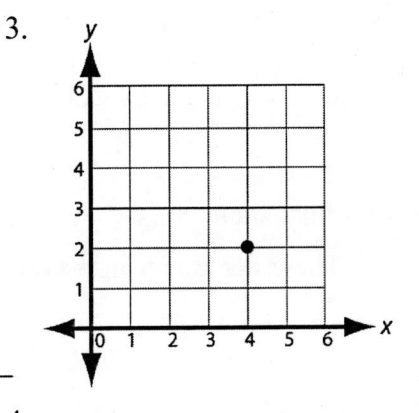

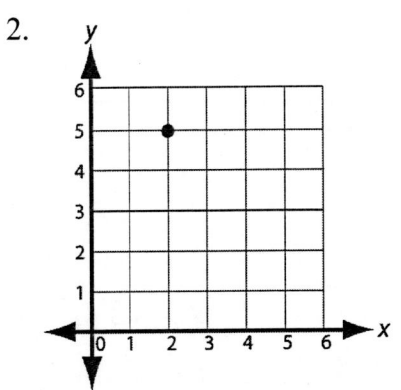

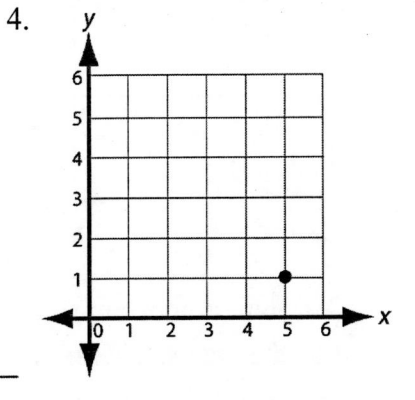

1.

2.

3.

4.

11.2 Plotting Points on a Coordinate Grid (DOK 1)

When given an ordered pair to plot, find the number on the x-axis first, then count up to the number on the y-axis.

Example 3: Find the point where the ordered pair $(1, 3)$ would be placed on the coordinate grid below.

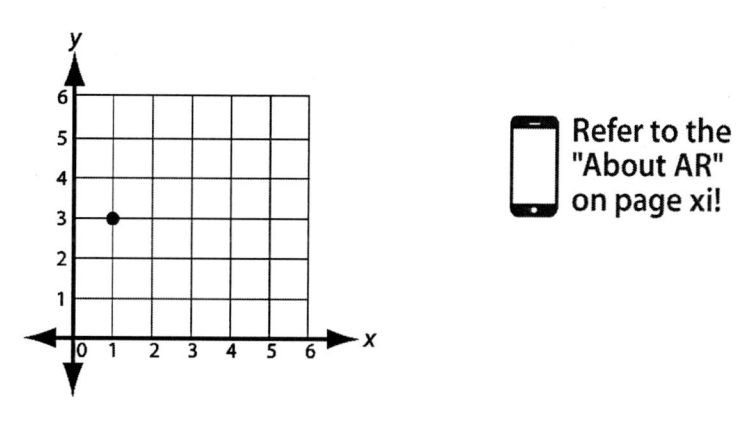

Refer to the "About AR" on page xi!

The number in the ordered pair for the x-axis is 1. So first, go to the number 1 on the x-axis. The number in the ordered pair for the y-axis is 3, so count up from the 1 to the third line. This is where you place the point for the ordered pair $(1, 3)$.

Plot the following ordered pairs. (DOK 1)

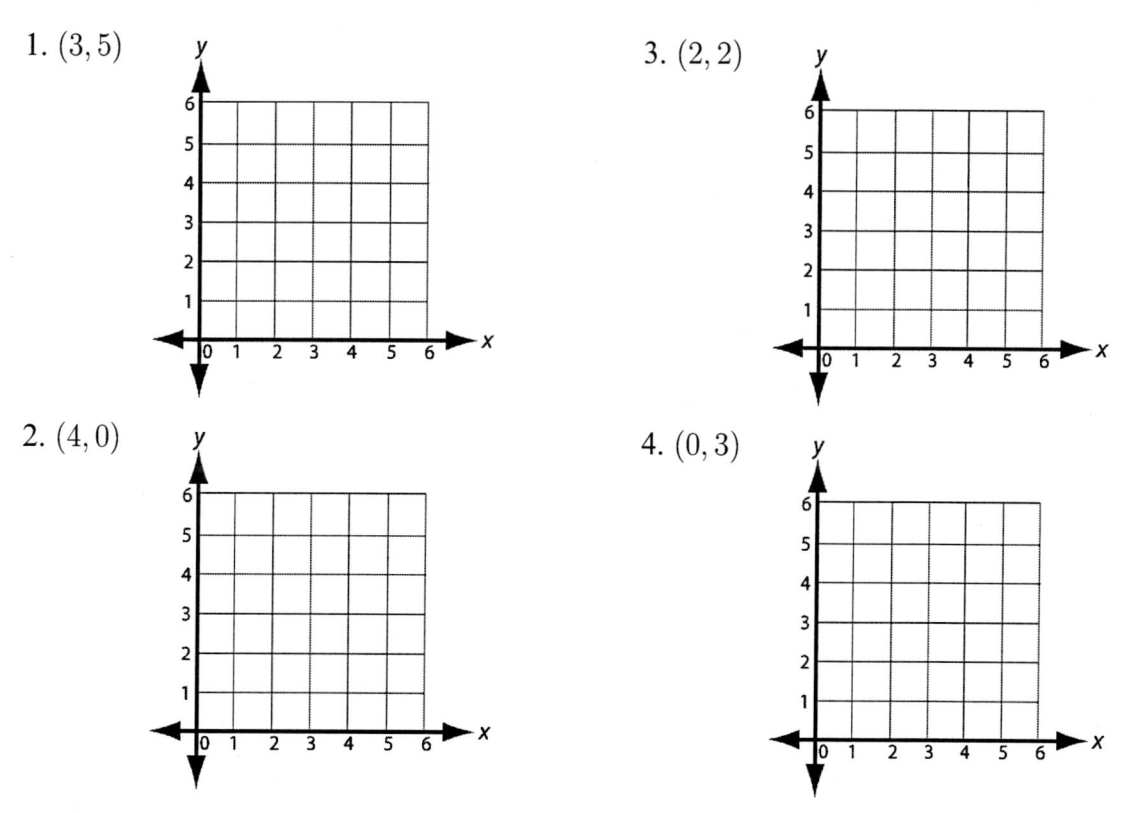

1. $(3, 5)$

2. $(4, 0)$

3. $(2, 2)$

4. $(0, 3)$

Chapter 11   Introduction to Graphing

5. (5, 4)

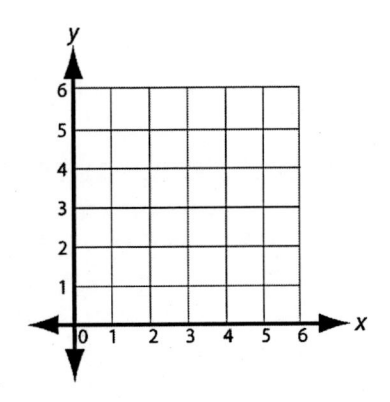

6. (1, 6)

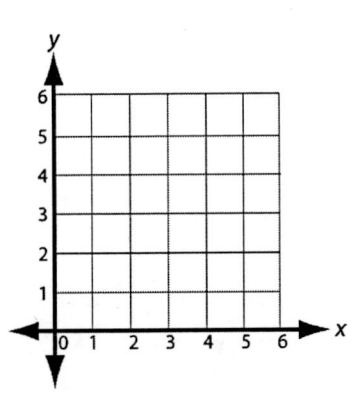

7. (2, 3)

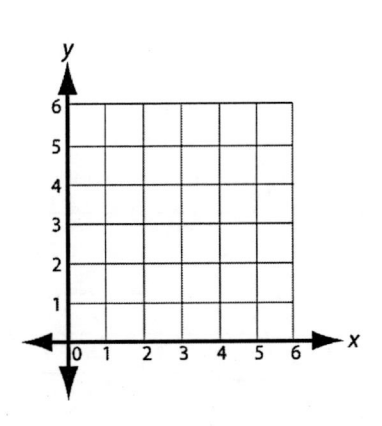

8. (3, 1)

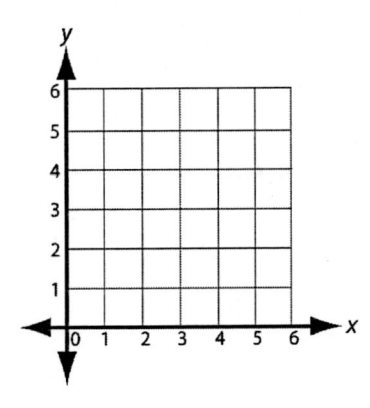

9. (0, 1)

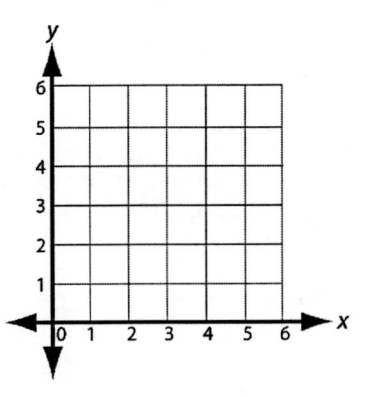

10. (6, 4)

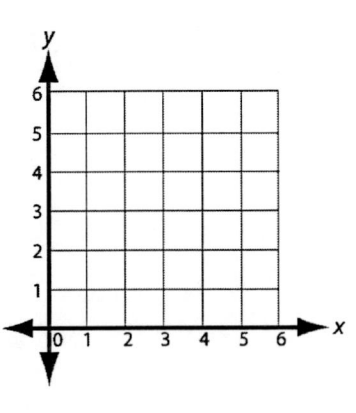

11. (4, 5)

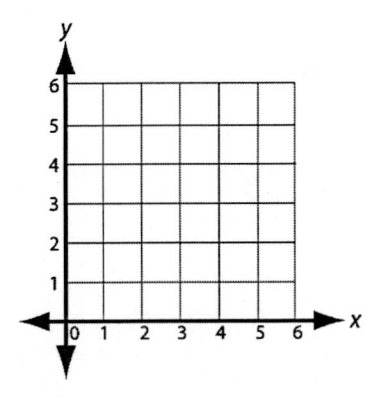

12. (3, 0)

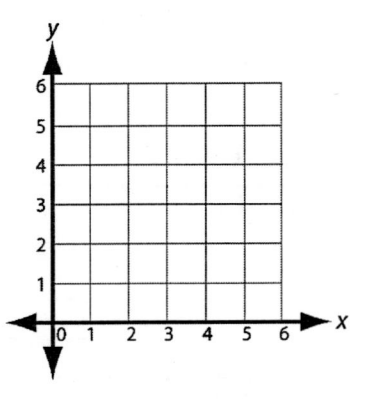

11.3 Finding Points From Diagrams (DOK 2, 3)

Diagrams often use a coordinate grid to plot locations on a map and designs for needlework and other pattern designs. The x and y axes usually use numbers on the y-axis and letters on the x-axis.

Use the graphic below for questions 1–3. (DOK 2)

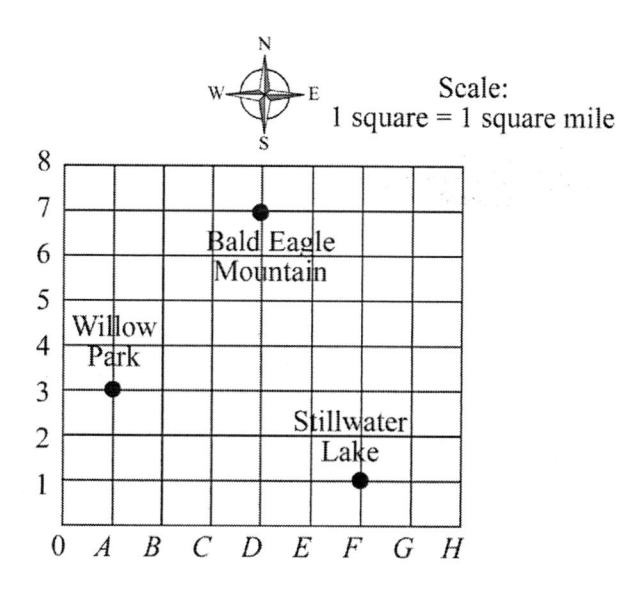

1. At what point is Willow Park?

2. What is at point $(D, 7)$?

3. At what point is Stillwater Lake?

Use the graphic below for questions 4–6. (DOK 2)

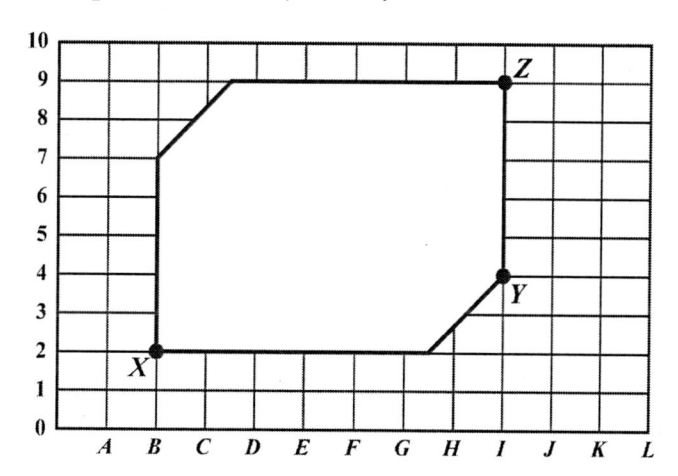

4. At what point is the corner labeled Z?

5. At what point is the corner labeled X?

6. At what point is the corner labeled Y?

Chapter 11   Introduction to Graphing

Use the graphic below for questions 7–9. (DOK 2, 3)

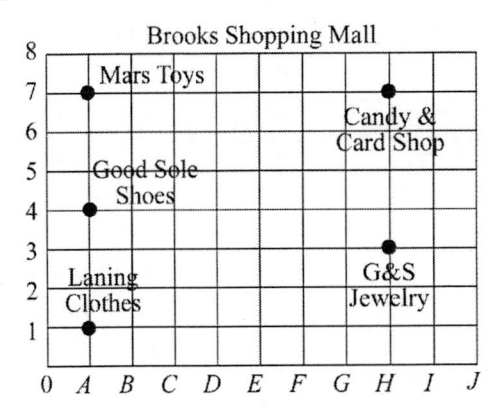

Brooks Shopping Mall — Mars Toys at (A, 7); Good Sole Shoes at (A, 4); Laning Clothes at (A, 1); Candy & Card Shop at (H, 7); G&S Jewelry at (H, 3).

7. Mrs. MacDougle was shopping with her daughter at the G&S Jewelry store. Next, they went to the Good Sole Shoes store. What are the coordinates for the Good Sole Shoes store?

8. How many spaces up and over did they travel to get to the Good Sole Shoes store from G&S Jewelry, according to the map?

9. Next, Mrs. MacDougle and her daughter went to the Candy & Card Shop. What are the coordinates for the Candy & Card Shop?

Use the graphic below for questions 10–11. (DOK 2, 3)

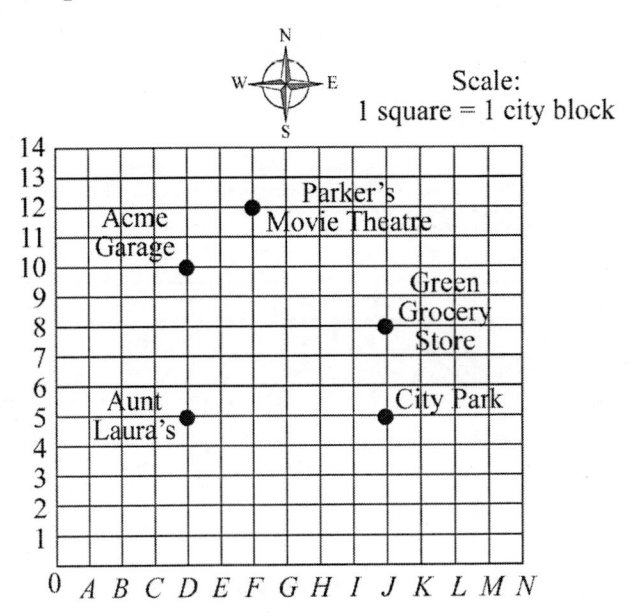

Scale: 1 square = 1 city block

Acme Garage at (C, 10); Parker's Movie Theatre at (E, 12); Green Grocery Store at (J, 8); Aunt Laura's at (C, 5); City Park at (J, 5).

10. The map above shows part of the town where James lives. If James leaves his Aunt Laura's house to go to the City Park, how many spaces to the right will he go according to the map?

11. What are the coordinates of the City Park?

12. When James leaves the City Park, he goes to meet his father at the Acme Garage. How many spaces to the left and towards the top does James go to get to the Acme Garage?

Chapter 11 Review

Locate the point in the coordinate grids below, and name the ordered pair. (DOK 1)

1. (2, 4) _____

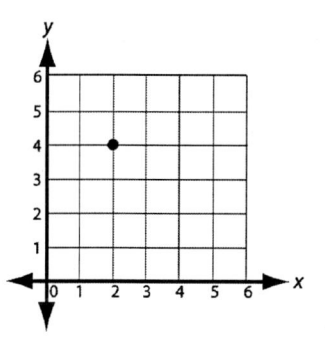

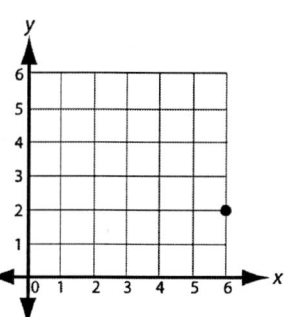

2. (6, 2) _____

3. (3, 3) _____

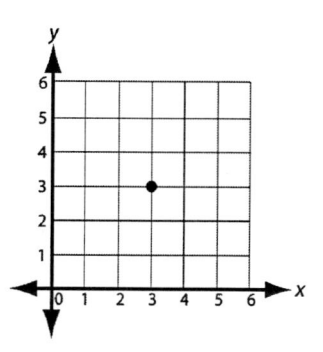

4. (5, 1) _____

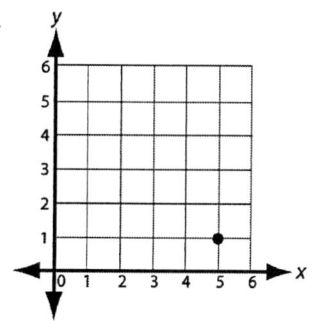

Plot the following ordered pairs. (DOK 1)

5. (6, 6)

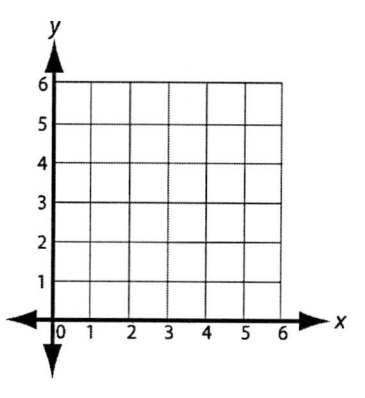

6. (0, 4)

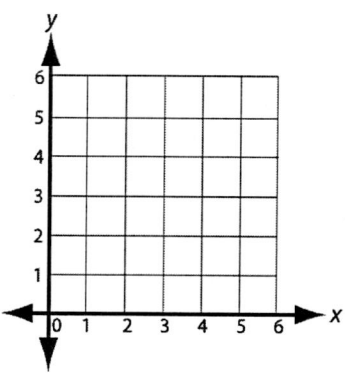

7. (5, 2)

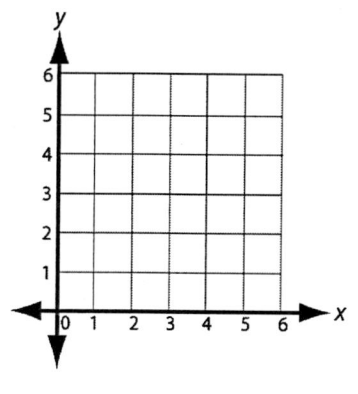

8. (4, 3)

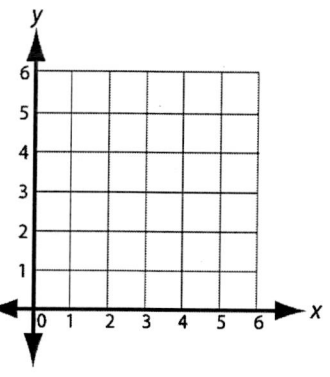

Chapter 11   Introduction to Graphing

Use the graphic below for questions 9–11. (DOK 2, 3)

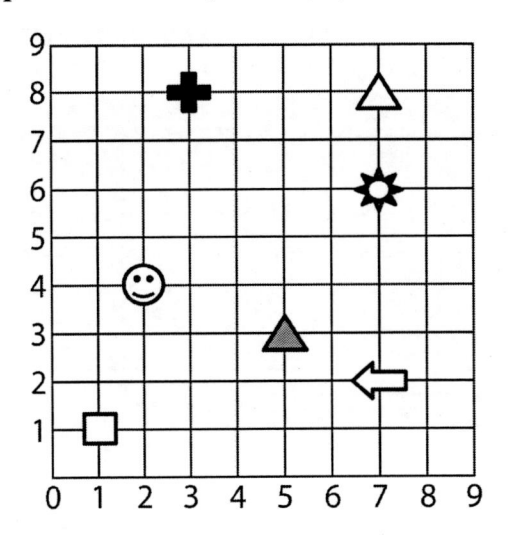

9. What are the coordinates for the happy face?

10. What are the coordinates for the sun?

11. Suppose the graphic above were part of a game board and you were the left pointing arrow. Staying on the grid lines, how many spaces up and over would you move to get to the black plus sign?

Use the graphic below for questions 12–14. (DOK 2, 3)

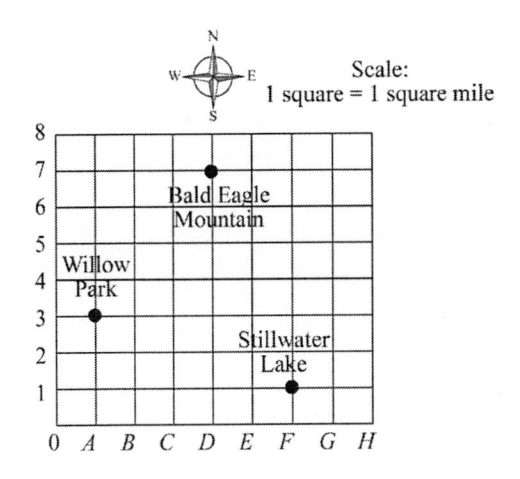

12. Robert and his scout troop are camping at Bald Eagle Mountain. The troop master decides to take the troop to Stillwater Lake. Using the key in the upper right corner and staying on the grid lines, how many miles south and to the east will the troop travel?

13. The troop changes location on the third day of the camping trip to Willow Park. What are the coordinates of Willow Park?

14. The troop stops to have lunch on coordinate $(C, 1)$. Staying on the grid lines, how many miles will they still need to travel north and west to get to Willow Park?

Chapter 11 Test

1 Which statement is true?

  A The x-axis is the vertical axis.
  B The y-axis is the vertical axis.
  C The x-axis is parallel to the y-axis.
  D The y-axis is the horizontal axis.

  (DOK 1)

2 Locate the point in the coordinate grid below and name the ordered pair.

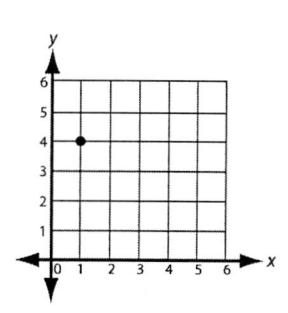

  A $(1, 0)$
  B $(0, 4)$
  C $(1, 4)$
  D $(4, 1)$

  (DOK 1)

3 In the ordered pair $(8, 3)$, which number is the y-coordinate?

  A 8
  B 0
  C 11
  D 3

  (DOK 1)

4 Locate the point in the coordinate grid below and name the ordered pair.

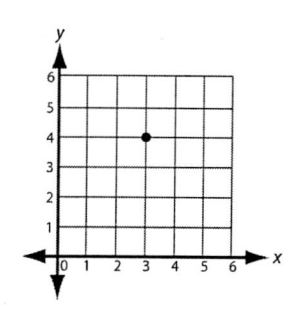

  A $(3, 0)$
  B $(0, 4)$
  C $(3, 4)$
  D $(4, 3)$

  (DOK 1)

5 In the ordered pair $(7, 2)$, which number is the x-coordinate?

  A 7
  B 0
  C 2
  D 9

  (DOK 1)

6 Locate the point in the coordinate grid below and name the ordered pair.

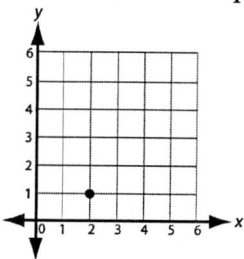

  A $(1, 2)$
  B $(0, 2)$
  C $(1, 0)$
  D $(2, 1)$

  (DOK 1)

7 Which statement is true?

  A The y-axis is the horizontal axis.
  B The y-axis runs parallel to the x-axis.
  C The x-axis is the horizontal axis.
  D The x-axis is the vertical axis.

  (DOK 1)

8 In the ordered pair $(3, 5)$, which number is the y-coordinate?

  A 5
  B 3
  C 8
  D 2

  (DOK 1)

Chapter 11   Introduction to Graphing

Locate the points in the coordinate grids below, and name the ordered pairs.

9

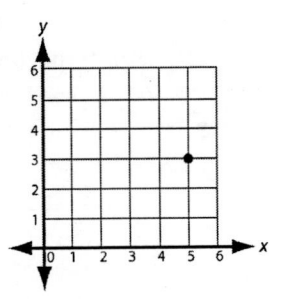

A (5, 0)
B (0, 3)
C (5, 3)
D (3, 5)

10

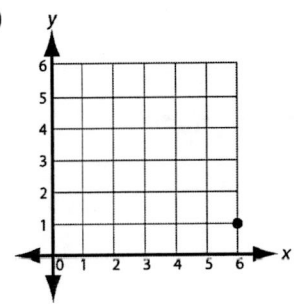

A (6, 0)
B (6, 1)
C (1, 6)
D (0, 1)

(DOK 1)

11

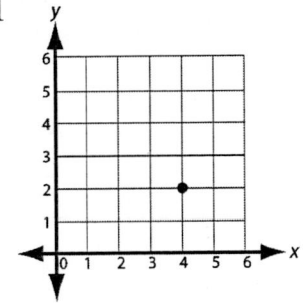

A (4, 0)
B (2, 4)
C (4, 2)
D (0, 4)

(DOK 1)

12

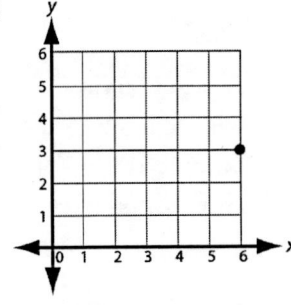

A (6, 0)
B (0, 3)
C (3, 6)
D (6, 3)

(DOK 1)

13

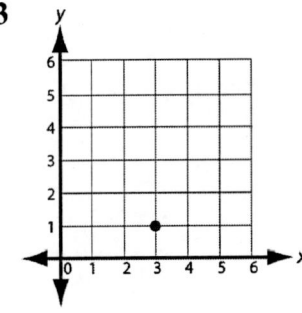

A (3, 0)
B (3, 1)
C (1, 3)
D (0, 1)

(DOK 1)

14

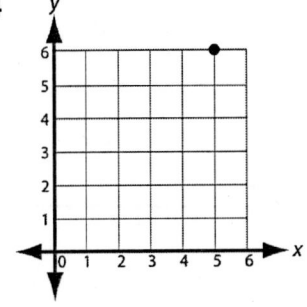

A (5, 6)
B (6, 5)
C (5, 0)
D (0, 6)

(DOK 1)

Chapter 11 Test

Use the graphic below to answer questions 15–18.

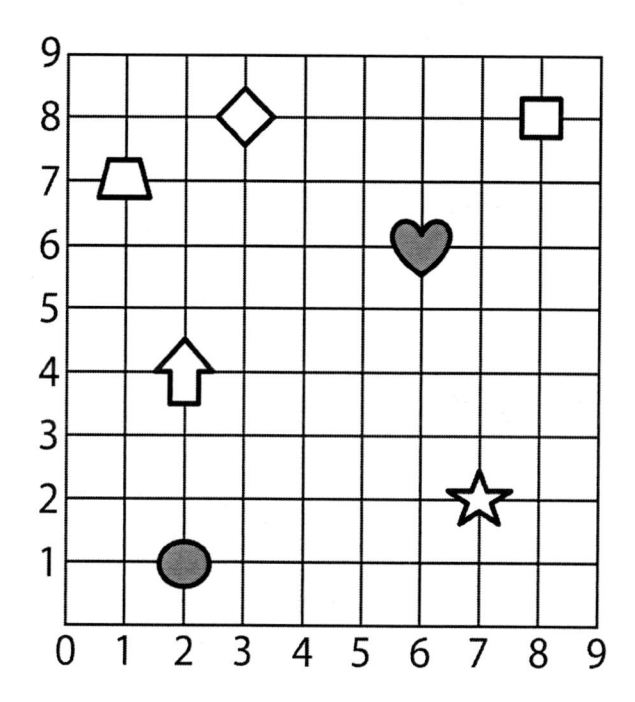

15 What are the coordinates for the heart?

A (0, 6)

B (6, 5)

C (6, 0)

D (6, 6)

(DOK 1)

16 Imagine the graphic is part of a game board and you are on the circle. Staying on the gridlines, how many steps up and to the right must you go to get to the star?

A 1 up and 4 to the right

B 2 up and 6 to the right

C 1 up and 7 to the right

D 1 up and 5 to the right

(DOK 2)

17 What are the coordinates for the trapezoid?

A (7, 1)

B (1, 7)

C (7, 0)

D (0, 7)

(DOK 1)

18 Imagine the graphic is part of a game board and you are on the up arrow. Staying on the gridlines, how many steps up and to the right must you go to get to the heart?

A 1 up and 4 to the right

B 2 up and 4 to the right

C 1 up and 6 to the right

D 2 up and 5 to the right

(DOK 2)

Copyright © American Book Company

Chapter 12
Algebra and Patterns

This chapter covers the following CC 5 standards:

Operations and Algebraic Thinking	5.OA.1, 5.OA.2, 5.OA.3

12.1 Algebra Vocabulary (DOK 1)

Vocabulary Words	Example	Definition
variable	$4x$ (x is the variable)	a letter that can be replaced by a number
coefficient	$4x$ (4 is the coefficient)	a number multiplied by a variable or variables
term	$5x + 3x - 6$ ($5x, 3x$, and -6 are terms)	numbers or variables separated by $+$ or $-$ signs
constant	$5x + x + 6$ (6 is a constant)	a term that does not have a variable
sentence	$2x = 7$ or $5 \leq x$	two algebraic expressions connected by $=, \neq, <, >, \leq, \geq,$ or $\approx$
equation	$4x = 8$	a sentence with an equal sign
inequality	$7x < 30$ or $x \neq 6$	a sentence with one of the following signs: $=, \neq, <, >, \leq, \geq,$ or $\approx$
parentheses, brackets, and braces	( ), { }, [ ] $[7 + \{8 \times (5 - 3)\}]$	used to enclose multiple steps in a number sentence

12.2 Understanding Algebra Word Problems (DOK 2)

Algebra word problems are used to describe real life situations and solve real life problems. The key to correctly solving a word problem is to correctly express the verbal ideas in algebraic form. This is where the word "expression" gets its meaning in algebra. A simple list of what to do will help you solve algebra word problems. Vocabulary will help you determine which operation should be completed to correctly answer the problem.

Do

1. Read the problem CAREFULLY.
2. Decide what is known. These terms are expressed as constants (integers).
3. Decide what is unknown. These terms are expressed as variables. (Whether or not the variable has a coefficient is based on each problem.)
4. Decide what the question is asking. Determine the operation needed to be done.

The list of keywords below will help you identify the operation needed to be done.

Operation	Keywords
Addition	increased by, is added, totals, combined, more than, plus, more, sum, and
Subtraction	difference, subtracted from, subtracted by, decreased by, minus, less, lower than
Multiplication	times, each, of, double, twice, half, multiply, product, triple
Division	divided by, quotient, divided into, divided among, ratio of/to

Use what you have learned to match each verbal expression to the letter of its matching algebraic expression. Some letters will be used more than once. (DOK 2)

1. four less than a number   A. $x - 4$

2. a number multiplied by four

3. a number added to four   B. $x + 4$

4. a number divided by four

5. four times a number   C. $\dfrac{x}{4}$

6. x increased by four

7. a number decreased by 4
                                          D. $4x$
8. one fourth of a number

Chapter 12 Algebra and Patterns

Write an algebraic expression for each word problem below. (DOK 2)

9. five guests more than planned, p

10. the class, c, with eight students missing

11. a number, n, decreased by thirty-one

12. the difference of a number, n, and eighteen

13. the sum of four times a number, n, and six

14. the product of eight and three times a number, n

15. nine dollars minus purchases, p

16. a number, n, times 0.8

17. the total number of cupcakes, c, divided among four trays

18. half the number of cookies, c, plus seven extra

19. bacteria culture, b, doubled

20. triple John's age, y

21. n feet lower than 10

22. 3 more than p

23. the product of 4 and m

24. a number, y, decreased by 20

25. 5 times as much as x

26. a number, n, plus 4

27. quantity, t, less 6

28. 18 divided by a number, x

12.3 Equivalent Expressions (DOK 2)

Example 1: Mavis has purchased 12 apples, three times in the last month. Which is the equivalent expression to this sentence?

A) $12 + 3$ B) $12 - 3$ C) 12×3 D) $12 \div 3$

Step 1: Ask yourself what the sentence is talking about. In this case, Mavis has purchased 12 apples on 3 different occasions. How many apples is this?

Step 2: Choose the answer that fits the problem. C is the answer. Mavis has purchased 12 apples times 3 occasions.

Answer: C

Choose the expression from the column on the left for each expression on the right. (DOK 2)

1. Which expression shows a way to compute 37×24?

2. Which expression is equivalent to $42 + 37$?

3. Sophia sells necklaces she makes for $24 each. Every week she makes and sells 20 necklaces. Which expression shows how much she sells all of the necklaces for in n weeks?

4. A box of shirts will hold 24 shirts. Which expression shows n boxes of shirts?

5. Which expression shows a way to compute $420 \div 20$?

6. Which expression shows how Alan may calculate the annual total of his weekly allowance, n?

7. Miss Jackson needs to calculate how many jelly beans to buy for her class of 24 students, so each student gets the same number of jelly beans, b.

8. Which expression is equivalent to 16 classes $\times$ s students?

9. Which expression shows a way to compute 24×22?

10. There are b number of bees in a bee hive. Mr. Bumble has 24 hives on his farm, half with the same number of bees, and half with $b - 10$ number of bees. Which expression shows the total number of Mr. Bumble's bees?

A) $(24 \times 20) \times n$

B) $52 \times n$

C) $(400 \div 20) + (20 \div 20)$

D) $(12 \times b) + 12(b - 10)$

E) $16 \times s$

F) $(20 \times 22) + (4 \times 22)$

G) $n \times 24$

H) $24 \times b$

I) $37 + 42$

J) $(30 \times 24) + (7 \times 24)$

Chapter 12   Algebra and Patterns

Choose the equivalent expression from the column on the right for each expression on the left. (DOK 2)

11. Which expression is equivalent to 84 less 24?

12. Which expression shows a way to compute 3 times 7 plus 3 times 11?

13. Roger mows lawns for $20 each. If he mows 8 lawns every week during summer break, a total of 11 weeks, how much money will Roger make?

14. Which expression is equivalent to x minus 60?

15. Which expression shows a way to compute $99 \div 9$?

16. Which expression shows a way to compute 24×84?

17. Which expression is equivalent to x times 84?

18. Mr. Walker wants to figure out how many hotdogs to buy for a family reunion of 24 people. If each person eats 2 hotdogs, which expression shows the total number of hotdogs needed?

19. Which expression shows how Justin may calculate the total of his birthday gifts, $5 from each of 6 relatives?

20. Amy walks about 5 miles going back and forth to school each week. If there are 36 school weeks, which expression shows the number of miles Amy will walk back and forth in a school year?

K) ($20 \times 8) \times 11$

L) $(90 \div 9) + (9 \div 9)$

M) 24×2

N) $(3 \times 7) + (3 \times 11)$

O) 36×5

P) $84x$

Q) $(24 \times 80) + (24 \times 4)$

R) $84 - 24$

S) $6 \times \$5$

T) $x - 60$

12.4 Evaluating Expressions (DOK 2)

There are some number expressions that require many steps to solve. Steps must be done starting with the innermost expression.

Example 2: Solve: $[3 + \{50 - 2(4 + 3)\}]$

Rule #1: First solve the problem within the **parentheses**. $(4 + 3) = 7$

Rule #2: Next, solve the problem within the **brackets**. $\{50 - 2(4 + 3)\} = \{50 - 2(7)\} = 50 - 14 = 36$

Rule #3: Lastly, solve the problem within the **braces**. $[3 + 36] = 39$
We know from step #2 that everything inside the brackets $= 36$.

Answer: 39

Example 3: Solve: $2(3 + 5)$. Solve: $2(3 + 5)$.
Done <u>incorrectly</u>: Done <u>correctly</u>:

Solution: $2 \times 3 = 6 + 5 = 11$ $3 + 5 = 8 \times 2 = 16$

In summary:
1st: Do the math inside the parentheses. ( )
2nd: Do the math inside the brackets. { }
3rd: Do the math inside the brackets. [ ]

Solve the expressions following the rules in Example #2. (DOK 2)

1. $[7 + \{18 + (7 - 2)\}]$
2. $[40 \div \{4(6 - 4)\}]$
3. $[8 \times \{16 - (23 - 17)\}]$
4. $[5 + \{4 \times (10 - 8)\}]$
5. $[36 \div \{12 \div (9 - 3)\}]$
6. $[22 - \{7 \times (5 - 3)\}]$
7. $[43 + \{90 \div (15 - 12)\}]$
8. $[28 \div \{3 + (16 \div 4)\}]$
9. $[100 - \{100 - (100 - 50)\}]$
10. $[3 \times \{63 \div (3 \times 3)\}]$
11. $[72 - \{30 \div (18 \div 3)\}]$
12. $[25 + \{25 + (25 + 25)\}]$
13. $[48 + \{12(6 + 2)\}]$
14. $[700 - \{18 \div (4 + 5)\}]$
15. $[222 + \{22 + (2 \times 2)\}]$
16. $[52 - \{6(10 - 5)\}]$
17. $[125 + \{25(6 - 4)\}]$
18. $[11\{3(9 - 5)\}]$

Chapter 12   Algebra and Patterns

12.5   Number Patterns (DOK 3)

Number **patterns** using addition and subtraction are defined by math equations. For instance, counting by two's is a number pattern using addition. Or starting from 100 and counting backwards by fours is a number pattern using subtraction.

Example 4:   Find the pattern and the next number in the pattern. 1, 3, 5, 7, ___?

   Step 1:   Find the pattern. The numbers are increasing by 2.

   Step 2:   Find the next number. $7 + 2 = 9$. The next number in the pattern is 9.
   Answer: +2; 9

Example 5:   Find the pattern and the missing number in the pattern. 30, 25, 20, ___, 10, 5.

   Step 1:   Find the pattern. The numbers are decreasing by 5.

   Step 2:   Find the next number. $20 - 5 = 15$. The number missing in the pattern is 15.
   Answer: −5; 15

Find the pattern and then the missing number(s) in the following patterns using addition and subtraction. (DOK 3)

1. 2, 4, 6, ___, 10, 12, ___, ___

2. 100, 90, 80, 70, ___, ___, ___

3. 60, 63, 66, ___, 72, 75, ___, ___

4. 17, 15, 13, 11, ___, 7, ___, ___

5. 40, 44, ___, 52, 56, 60, ___, ___

6. 18, 23, 28, ___, 38, 43, ___, ___

7. 94, 88, 82, 76, ___, 64, ___, ___

8. 11, 19, ___, 35, 43, 51, ___, ___

9. 62, 59, 56, 53, ___, 47, ___, ___

10. 55, 53, 51, ___, 47, 45, ___, ___

11. 32, 42, ___, 62, 72, 82, ___, ___

12. 300, 250, 200, ___, 100, ___, ___

13. 55, 66, ___, 88, 99, 110, ___, ___

14. 220, 210, 200, 190, ___, ___, ___

15. 717, 723, 729, ___, 741, ___, ___

16. 600, 575, 550, ___, 500, ___, ___

12.5 Number Patterns (DOK 3)

Number patterns using multiplication and division are a little more challenging. If the pattern isn't obvious, you will need to figure it out by trial and error. There are clues you can use to determine if the pattern is using multiplication or division. If the pattern is increasing, you will use multiplication. If the pattern is decreasing, you will use division.

Example 6: Find the pattern and the next number in the pattern. 2, 6, 18, 54, ___ ?

Step 1: Find the pattern. The numbers are increasing by a factor of 3.

Step 2: Find the next number. The next number in the pattern would be $54 \times 3 = 162$.

Answer: ×3; 162

Example 7: What number is missing in the pattern 1000, 100, ___, 1?

Step 1: Find the pattern. The numbers are decreasing by a divisor of 10.

Step 2: Find the missing number. The missing number in the pattern is $100 \div 10 = 10$.

Answer: ÷10; 10

Find the pattern and the missing number(s) in the following patterns using multiplication and division. (DOK 3)

1. 125, 25, ___, 1

2. 2, 4, 8, ___, 32, 64, ___, ___

3. 80, 40, 20, ___, 5

4. 3, 12, ___, 192, 768, ___, ___

5. 640, 80, ___

6. 11, ___, 1331, 14641, ___, ___

7. 1000, 500, ___, 125, ___

8. 200, 400, 800, ___, ___, ___

9. 3, 6, 12, ___, 48, 96, ___, ___

10. 4, 12, ___, 108, 324, ___, ___

11. 6, 12, 24, ___, 96, ___, ___

12. 640, 160, ___, 10

13. 1, 7, ___, 343, 2401, ___, ___

14. 240, 120, ___, 30, 15

15. 3, 9, 27, ___, 243, ___, ___

16. 1, 1, 1, ___, 1, 1, 1, ___, ___

Chapter 12 Algebra and Patterns

12.6 Pattern Rules (DOK 2, 3)

Example 8: Given the rule "+ 2" and starting with zero, give the next three numbers for the rule.

Directions: Start with zero and add 2. Do this three times.
0, 2, 4, 6

Answer: 0, 2, 4, 6

Example 9: Given the rule "÷ 3" and starting with 270, give the next three numbers for the rule.

Directions: Start with 270 and divide by 3. Do this three times.
270, 90, 30, 10

Answer: 270, 90, 30, 10

Follow the directions for the rule given and the number to start with for each problem. Give the next three numbers after the number given. The first one has been done for you. (DOK 2)

1. Rule: − 7
   Start with 80.
   80, 73, 66, 59

2. Rule: × 2
   Start with 1.

3. Rule: + 18
   Start with 0.

4. Rule: ÷ 2
   Start with 100.

5. Rule: + 3
   Start with 15.

6. Rule: − 5
   Start with 75.

7. Rule: + 4
   Start with 20.

8. Rule: − 11
   Start with 88.

9. Rule: ÷ 5
   Start with 5,000.

10. Rule: + 6
    Start with 10.

11. Rule: − 2
    Start with 22.

12. Rule: + 14
    Start with 0.

13. Rule: × 4
    Start with 2.

14. Rule: + 10
    Start with 100.

15. Rule: − 15
    Start with 75.

16. Rule: × 3
    Start with 1.

17. Rule: − 20
    Start with 100.

18. Rule: + 100
    Start with 37.

12.6 Pattern Rules (DOK 2, 3)

Example 10: Given the rule "+ 4" and starting with 0, give the next three numbers for the rule. And, given the rule "+ 8" and starting with 0, give the next three numbers for the rule. Lastly, compare the two sets of numbers.

Step 1: Make a chart of the two rules and fill the chart in.

Rule "+ 4"	Rule "+ 8"
0	0
4	8
8	16
12	24

Step 2: Compare the two sets of numbers. The rule for "+ 8" is twice as large as the rule for "+ 4".

Fill in the charts and give a short sentence of the different outcomes for each pair of rules. (DOK 3)

1. Start with zero.

Rule "+ 2"	Rule "+ 3"

2. Start with 100.

Rule "− 10"	Rule "− 5"

3. Start with two.

Rule "× 2"	Rule "× 4"

4. Start with 500.

Rule "÷ 5"	Rule "÷ 2"

5. Start with zero.

Rule "+ 10"	Rule "+ 100"

6. Start with 300.

Rule "− 50"	Rule "− 5"

7. Start with zero.

Rule "+ 10"	Rule "+ 15"

8. Start with 80.

Rule "− 2"	Rule "− 4"

9. Start with 60.

Rule "+ 20"	Rule "+ 40"

Copyright © American Book Company

Chapter 12 Algebra and Patterns

12.7 Graphing Patterns (DOK 2)

Example 11: Fill in the chart for the two rules below, starting with one. Then graph the results on a coordinate grid.

Step 1:

x-axis	y-axis
Rule "+ 2"	Rule "+ 3"

x-axis	y-axis
Rule "+ 2"	Rule "+ 3"
1	1
3	4
5	7
7	10

Step 2: Graph the points on the coordinate grid.

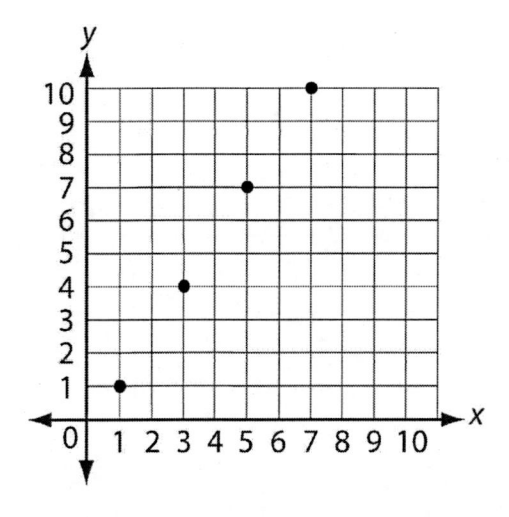

Fill in the charts and graph the points on a coordinate grid. (DOK 2)

1. Start with 1.

x-axis	y-axis
Rule "+ 1"	Rule "+ 2"
1	1

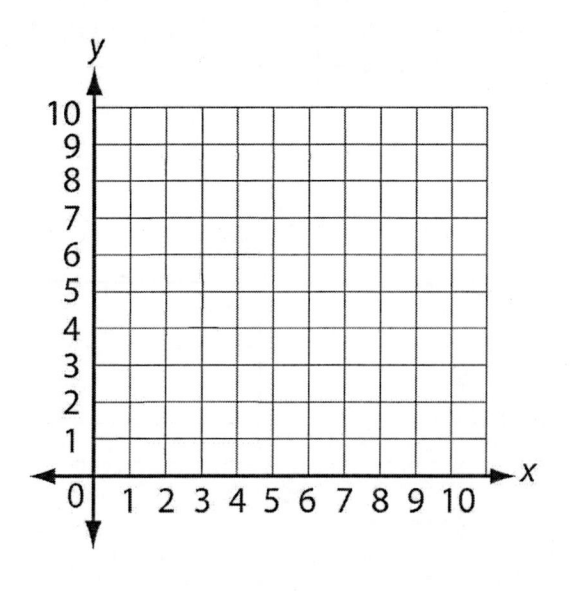

164   Copyright © American Book Company

12.7 Graphing Patterns (DOK 2)

2. Start with 10.

x-axis	y-axis
Rule "− 2"	Rule "− 1"
10	10

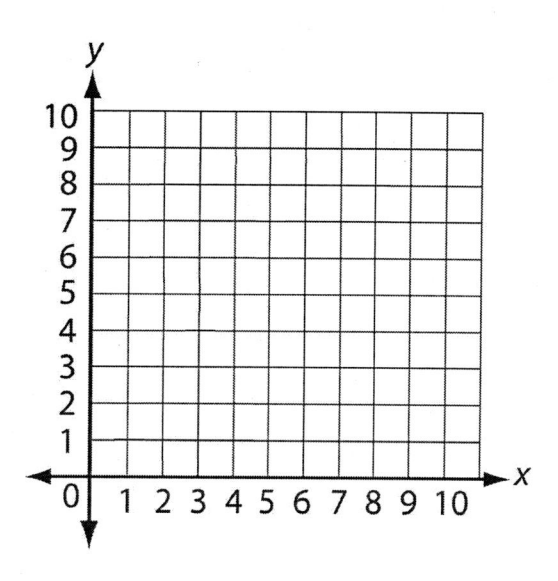

3. Start with 1.

x-axis	y-axis
Rule "× 1"	Rule "+ 2"
1	1

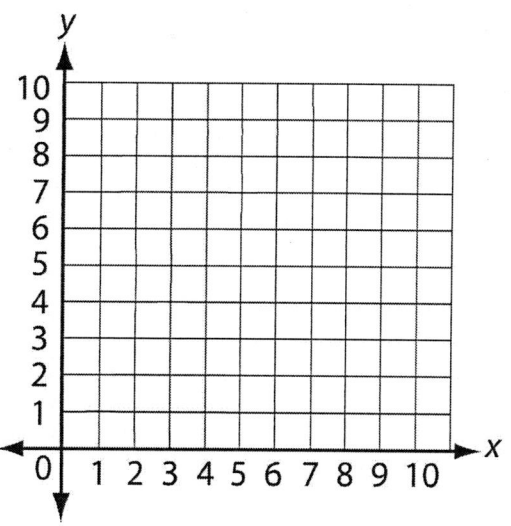

4. Start with 1.

x-axis	y-axis
Rule "+ 1"	Rule "× 2"
1	1

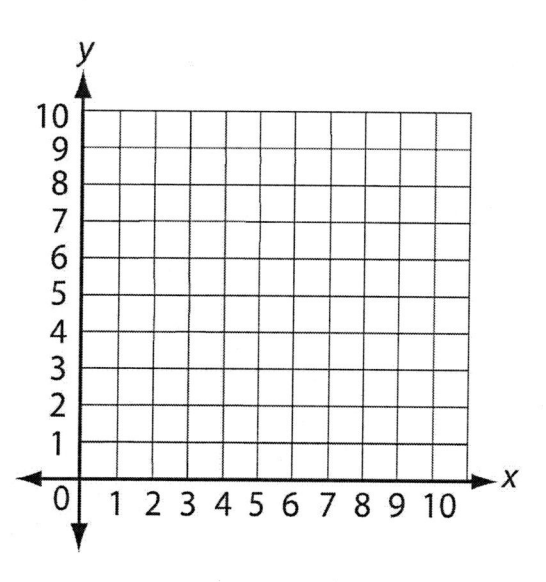

Chapter 12 Algebra and Patterns

12.8 Going Deeper into Algebra and Patterns (DOK 3)

Study each pattern, find the rule, <u>and</u> give the next three terms in the rule. (DOK 3)

1. ★   ★ ★   ★ ★ ★
              ★   ★
                         ★

2. 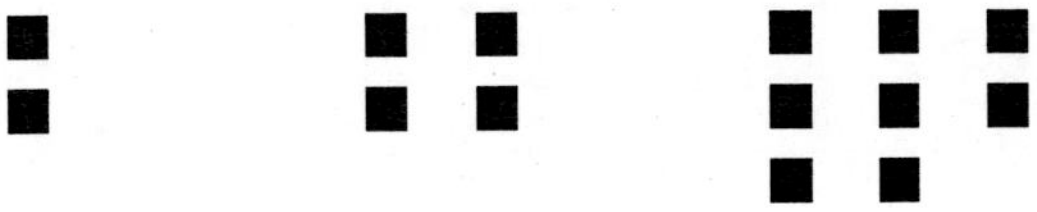 ■   ■ ■   ■ ■ ■
    ■   ■ ■   ■ ■ ■
                         ■ ■

3. Pattern: | 5 | 11 | 17 | 23 | 29 |

4. Pattern: | 1,000,000 | 200,000 | 40,000 | 8,000 |

5. Pattern: | 250 | 233 | 216 | 199 | 182 |

6.

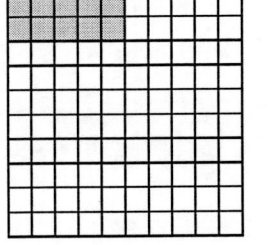

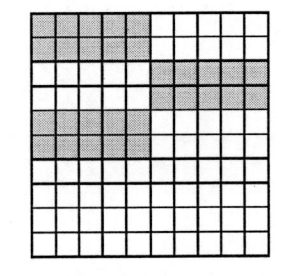

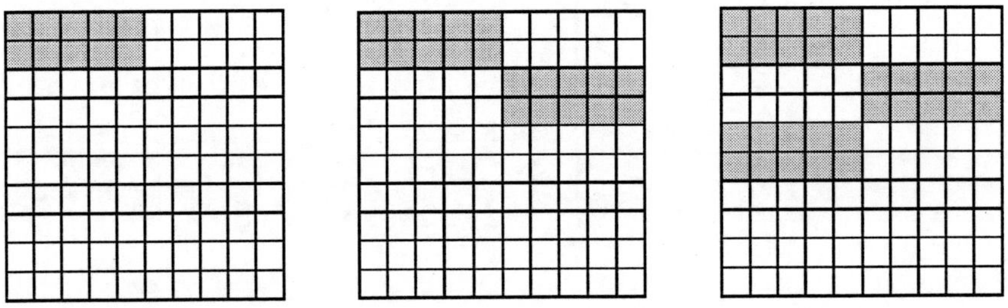

Use the grids below for your answer.

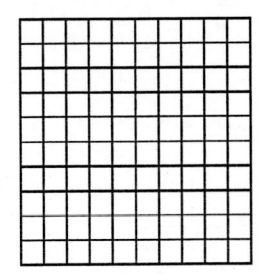

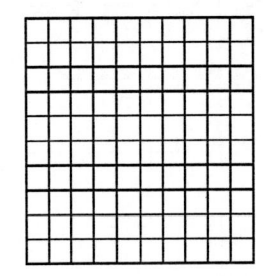

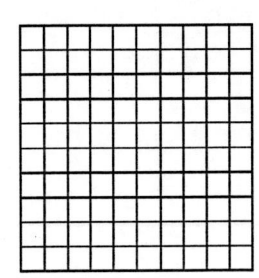

7. Alvin was given the pattern: | 176 | 88 | 44 | 22 | 11 |. He gave the answer for the rule as "-44×2". His answer is incorrect. What is the rule for the pattern and give the next three numbers in the pattern.

Chapter 12 Review

Write an algebraic expression for each verbal expression below. You do not have to solve the problem. (DOK 2)

1. a number less than seven

2. a number plus eight

3. three times a number

4. a number increased by sixteen

Write an expression for the following word problems. You do not have to solve the problem. (DOK 2)

5. Juan and Chandelle each step on a scale. Juan (j) weighs 6 pounds more than Chandelle. Write an expression to show the sum of the weights of the two boys.

6. A hamster weighs h ounces. A gerbil weighs 2 ounces more than a hamster. Write an expression of how much the two animals weigh together.

Choose the equivalent expression from the column on the right for each expression on the left. (DOK 2)

7. Which expression shows a way to compute 16×11?   A) $(10 + 6) + (10 + 1)$

8. Which expression is equivalent to $16 + 11$?   B) $16 - 11$

9. Rita has 16 seashells. She gives 11 of them to her cousin.   C) $(10 \times 11) + (6 \times 11)$

Fill in the charts and graph the points on a coordinate grid. (DOK 2)

10. Start with 1.

x-axis	y-axis
Rule "+ 1"	Rule "+ 3"
1	1

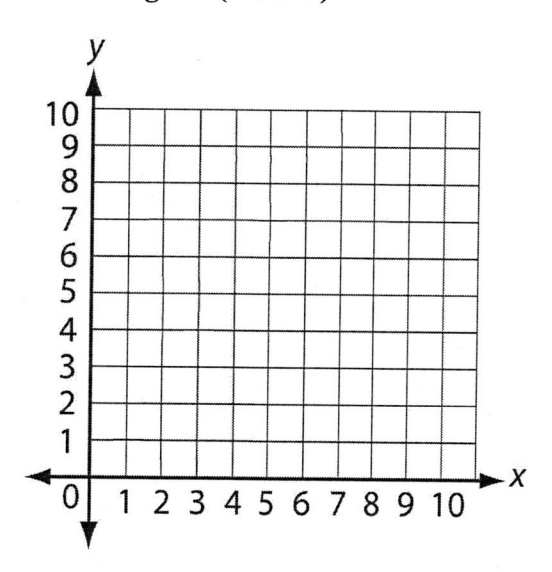

Chapter 12   Algebra and Patterns

Solve the expressions following the rules of parentheses, brackets, and braces. (DOK 2)

11. $[22 + \{90 - (85 - 15)\}]$

12. $[13 \times \{2(5 - 4)\}]$

13. $[100 \div \{36 - (5 + 6)\}]$

14. $[1,000 - \{4 \times (18 - 12)\}]$

Find the rule and then the missing number in the following patterns. (DOK 3)

15.   2, 5, 8, ___, 14, 17

16.   2, 20, 200, ___, 20,000

17.   25, 20, 15, ___, 5

Follow the directions for the rule given and the number to start with for each problem. Give the next three numbers after the number given. (DOK 2)

18. Rule: − 4

    Start with 60.

19. Rule: × 3

    Start with 1.

20. Rule: + 12

    Start with 0.

Fill in the charts and give a short sentence of the different outcomes for each pair of rules. (DOK 3)

21. Start with zero.

Rule "+ 3"	Rule "+ 6"

22. Start with 100.

Rule "− 5"	Rule "− 2"

Chapter 12 Test

1 Which algebraic expression is four more than a number?

- **A** $x - 4$
- **B** $x + 4$
- **C** $x \times 4$
- **D** $x \div 4$

(DOK 2)

2 Jacob has a plum, (p), and an apple that weighs twice as much as the plum. Which algebraic expression that shows the weight of both pieces of fruit?

- **A** $p + 2p$
- **B** $p - 2p$
- **C** $p + 2$
- **D** $p - 2$

(DOK 2)

3 Which expression is equivalent to n times (14 plus 7)?

- **A** $(n + 14) + (n + 7)$
- **B** $n(14 + 7)$
- **C** $n(14 - 7)$
- **D** $n(14 \times 7)$

(DOK 2)

4 Which expression is equivalent to 83×21?

- **A** $(80 \times 21) + (3 \times 21)$
- **B** $(83 \times 21) + (21 \times 83)$
- **C** $(83 + 21) \times (83 + 21)$
- **D** $(80 \times 21) \times (3 \times 21)$

(DOK 2)

5 Solve: $[48 - \{6 \times (2 \times 4)\}]$.

- **A** 18
- **B** 16
- **C** 2
- **D** 0

(DOK 2)

6 Solve: $[12 \div \{3 + (8 - 7)\}]$.

- **A** 1
- **B** 2
- **C** 3
- **D** 4

(DOK 2)

7 Which of the following charts shows the data in the coordinate grid below?

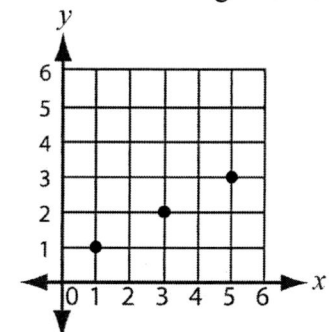

A

Rule "+1"	Rule "+2"
1	1
2	3
3	5

B

Rule "+2"	Rule "+2"
1	1
3	3
5	5

C

Rule "+2"	Rule "+1"
1	1
3	2
5	3

D

Rule "×1"	Rule "×2"
1	1
1	2
1	4

(DOK 2)

Chapter 12   Algebra and Patterns

8 Find the rule and the missing number in the pattern:
15, 30, 45, ___, 75.

A +15; 60

B +10; 55

C +20; 65

D −5; 70

(DOK 3)

9 Find the missing number in the pattern:
1, 3, 9, ___, 81.

A +3; 12

B ×3; 18

C ×3; 27

D ×4; 36

(DOK 3)

10 What are the first 4 numbers for the pattern rule "− 8", starting with 72?

A 72, 64, 56, 48

B 72, 64, 54, 44

C 72, 80, 88, 96

D 72, 80, 88, 98

(DOK 1)

11 What is the rule for the chart below?

Rule ?
1
7
49
343

A + 7

B × 7

C − 7

D ÷ 7

(DOK 2)

12 What are the first 4 numbers for the pattern rule "÷ 2", starting with 80?

A 80, 160, 320, 640

B 80, 60, 40, 20

C 80, 78, 76, 74

D 80, 40, 20, 10

(DOK 1)

13 Which of the following charts is filled out correctly? The charts should begin with the number 1.

A

Rule "+ 1"	Rule "+ 3"
1	1
2	3
3	5
4	7

B

Rule "× 2"	Rule "+ 2"
1	1
2	3
4	5
5	7

C

Rule "+ 3"	Rule "+ 6"
1	1
4	7
7	13
10	19

D

Rule "× 5"	Rule "× 10"
1	1
5	10
25	10
30	10

(DOK 2)

Chapter 13
How to Write Your Answers

13.1 Writing Short Answers

At times, you will need to do more than fill in blanks or make lists. Be sure to answer the question clearly. Look at this example below:

Example 1: Tammy needs $50 to go to her school dance, she has 5 family members willing to give her an equal amount of money so that she can go. How much money does each family member need to give Tammy?

This is a division problem so you would need to divide $50 papers by 5 family members: $50 ÷ 5 = $10

Answer: Each family member needs to give Tammy $10, so she can go to the dance.

13.2 Writing Open-Ended Answers

Read the Question Carefully

First of all, read the question carefully. Make sure you understand what it is asking. If you need help, ask your teacher.

Write Clearly

Answer each question in a clear way. If a question asks when something happened, be sure to talk about the sequence of events. If it asks you to compare two things, be sure to talk about how they are similar.

Being clear also means that your writing must be correct. Be sure to check your spelling. Also look for any other mistakes. If you find an error, erase it completely or cross it out. Write what you really wanted to say right above it or next to it.

Use Neat Handwriting

Make sure that the people reading your answer can tell what it is. Write in a neat way that others can read. Label your answer "answer:".

Chapter 13 How to Write Your Answers

Here is an example. Read the question carefully. Look at exactly what it asks you to do. Then, you will practice answering the question. You will also see how the best answer looks.

Example 2: Annie runs a soup kitchen on Saturdays in the inner city. In the last few years, as hard times have hit, Annie has noticed a greater use of her soup kitchen. The table below shows the number of people she feeds per week.

Year	1	2	3	4	5
# of people fed (per week)	50	65	80	95	?

A Describe how to find the next value in the table using words. Explain how you came to your answer.

B How many people will use Annie's Soup Kitchen per week in year 5, if the pattern continues?

C Write a number sentence using the letter n to represent the number of people fed in year 5.

Good Answer:

A Add 15 to the number of people fed each week. 95

B 110 people

C $95 + 15 = n$

Better Answer:

A Add 15 to the number of people fed each week. $50 + 15 = 65$; $65 + 15 = 80$; $80 + 15 = 95$

B $95 + 15 = 110$; 110 people

C Add 15 to the number of people fed in year 4 to find the number of people fed in year 5.
$95 + 15 = n$

Best Answer:

A Each year the number of people fed each week is increasing by 15.
Add 15 to the number of people fed each week to find the number of people fed each week in the next year.
$50 + 15 = 65$; $65 + 15 = 80$; $80 + 15 = 95$

B Add 15 to the number of people fed in year 4 to find the number of people fed in year 5.
$95 + 15 = 110$
If the pattern continues, the number of people fed in year 5 is 110 people.

C Let n represent the number of people fed in year 5.
Add 15 to the number of people fed in year 4 to find the number of people fed in year 5.
$95 + 15 = n$

Chapter 13 Review

You can practice writing some answers here. When you are done, think about if your answer is a good, better, or best answer for the question being asked. Review your answers with your teacher or tutor.

1 Use the map of the Brooks Shopping Mall to follow the path that Nellie May and her mother took while shopping one Saturday. Answer the four questions below the map.

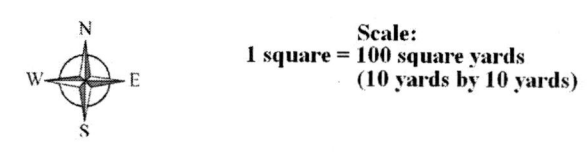

First, Nellie May and her mother went to the G&S Jewelry store.

Second, they went 40 yards north.

Thirdly, they went 70 yards west.

Finally, they went 30 yards south.

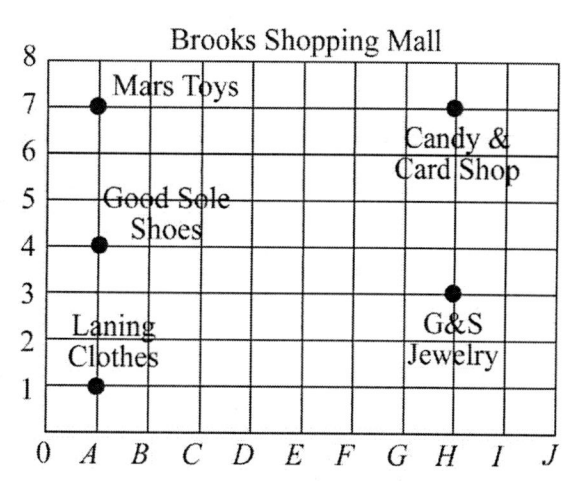

A Which store did they go to third?

B What is the name of the store they went to last?

C What is the coordinate pair on the map for the last store they went?

D Starting at the first store, how many yards did they walk from store to store until the last store they went?

Chapter 13   How to Write Your Answers

2 Using the survey chart below, answer the 3 questions that follow the chart.

Favorite Outdoor Games

	Soccer	Hopscotch	Playing Catch
Boys	40	0	35
Girls	30	25	15
Teachers	2	1	1

A   How many boys prefer to play catch?

B   What is the total number of boys, girls, and teachers who prefer hopscotch?

C   What is the total number of girls who answered the survey?

Index

Acknowledgements, ii
Algebra, 154
   coefficient, 154
   constant, 154
   equation, 154
   inequality, 154
   term, 154
   variable, 154
   word problems, 155
Area, 126
   rectangles, 126

Coefficient, 154
Constant, 154
Coordinate Grid, 143
   ordered pairs, 143
   plotting points, 145

Decimals
   adding decimals, 38
   division by decimals, 46
   division by whole numbers, 45
   expanded form, 30
   multiplication, 43
   ordering, 32
   reading and writing, 29
   rounding decimals, 31
   subtracting decimals, 39
Denominator, 54
Digits, 3
Divisibility, 15
Division
   reciprocals, 93
   remainders, 19

Equations, 154

Fact Families
   multiplication and division, 40
Fractions, 54
   adding, 60
   comparing, 68
   division, 91
   estimating, 66
   improper, 56
   least common multiple, 62
   lowest common denominator, 63
   mixed numbers, 65
      addition, 63
      subtraction, 64
   multiplying, 77, 82
   numerator, 58
   simplifying, 55
   simplifying improper fractions, 56
   simplifying with whole numbers, 57
   subtracting, 61

Geometry
   volume, 132

Improper Fractions
   simplifying, 56
Inequality, 154

Least Common Multiple, 62
Line Plots, 108
Lowest Common Denominator, 63

Measurement, 102
   customary, 102
   metric, 105
Mixed Numbers, 65
   addition, 63
   subtraction, 64
Multiplying
   by multiples of ten, 9

Number line, 143
Number problems, 160
Numerator, 54

Parallelogram, 120

Pattern Rules, 162
Place value, 1
Polygon, 118
Preface, vii

Quadrilateral, 119

Reciprocals, 93
Rectangle, 120
Rhombus, 120

Shapes
    catagories, 125
    properties of shapes
        angles and sides, 122
Square, 120

Table of Contents, vi
Term, 154

Variable, 154
Volume, 132
    compound figures, 135
    cube, 134
    rectangular prism, 133

Whole Numbers
    division, 17